Holt California Math

Focus on CAHSEE Standards: Intervention Workbook

HOLT, RINEHART AND WINSTON
A Harcourt Education Company
Orlando • Austin • New York • San Diego • London

Printed in the United States of America

ISBN: 978-0-03-099245-2
ISBN: 0-03-099245-1

4 5 862 11 10 09 08

CONTENTS

CONTENTS, *CONTINUED*

Name ______________________________ Date ____________ Class ____________

STANDARD 6PS1.1

Focus on CAHSEE Standards: Intervention

Sample Problems

Grade 6 • Statistics, Data Analysis, and Probability 1.1

Compute the ~~range~~, mean, median, and mode of data sets.

1. The table below shows the length of pencils in Sara's pencil box.

Length of Pencils						
5	4	3	7	4	4	8

Find the mean, median, and mode of the data set.

First put the numbers of the data set in numerical order, from least to greatest.

3, 4, 4, ___, ___, ___, ___

To find the mean, first find the sum of the numbers in the data set.

5 + 4 + 3 + ___ + ___ + ___ + ___ = ___

How many items are in the data set? ___

Divide the sum of the numbers in the data set by the number of items.

$$\frac{5+4+3+\square+\square+\square+\square}{\square} = \frac{\square}{\square} = \_\_\_$$

The mean is ___.

To find the median, find the value in the middle of the data set:

___ ___, ___, ___, ___, ___, ___

The median is ___.

The mode is the value that occurs most often.

The mode is ___.

2. The table below shows daily high temperatures for a week in January.

High Temperatures						
52	58	59	58	53	55	50

Find the mean, median, and mode of the data set.

First put the numbers of the data set in numerical order, from least to greatest.

___, ___, ___, ___, ___, ___, ___

To find the mean, first find the sum of the numbers in the data set.

___ + ___ + ___ + ___ + ___ + ___ + ___

= _____

How many items are in the data set? ___

Divide the sum of the data set by the number of items.

_____ ÷ ___ = ___

The mean is ___.

To find the median, find the value in the middle of the data set:

___, ___, ___, ___, ___, ___, ___

The median is ___.

The mode is the value that occurs most often.

The mode is ___.

Name ______________________ Date ____________ Class ____________

Focus on CAHSEE Standards: Intervention

STANDARD **6PS1.1** *Skills Practice*

Grade 6 • Statistics, Data Analysis, and Probability 1.1

Compute the ~~range~~, mean, median, and mode of data sets.

The table shows the number of hours Samantha studied each week.

Number of Hours				
10	14	15	11	10

Use the table to answer questions 1–3. Show your work.

1. What is the mean? ______

2. What is the median? ______

3. What is the mode? ______

The data below shows the amount of money Kalia collected for the soccer team two weeks in a row.

Week 1: \$11, \$13, \$18, \$20, \$16, \$9, \$18

Week 2: \$8, \$17, \$11, \$15, \$11, \$16, \$20

Use the data to answer questions 4–9. Show your work.

4. What is the mean for Week 1? ______

5. What is the mean for Week 2? ______

6. What is median for Week 1? ______

7. What is median for Week 2? ______

8. What is the mode for Week 1? ______

9. What is the mode for Week 2? ______

Name ______________________ Date __________ Class __________

STANDARD 6PS2.5

Focus on CAHSEE Standards: Intervention

Alternative Strategies

Grade 6 • Statistics, Data Analysis, and Probability 2.5

Identify claims based on statistical data and, in simple cases, evaluate the validity of claims.

Read the passage. Then use the checklist to answer the questions.

Alma took a random sample of 175 students in school, and she asked this question: "How much TV do you watch at night?" She concluded that 75% of students at school watch at least an hour of TV a night and that 25% watch more than two hours.

1. Complete the checklist:

a. What population is Alma studying?

____

b. Is a sample taken from this population? If so, what kind of sample?

____

____

c. If it's a not a random sample, is the sampling done correctly?

____

d. Is the survey question unbiased? ______

2. Make a circle graph showing Alma's conclusion.

3. Is Alma's conclusion valid? Explain.

____

____

____

Read the passage. Then use the checklist to answer the questions.

Lucy took a random sample of 500 students in school. She asked, "How much time do you waste in front of the TV?" She made a circle graph to show her survey results.

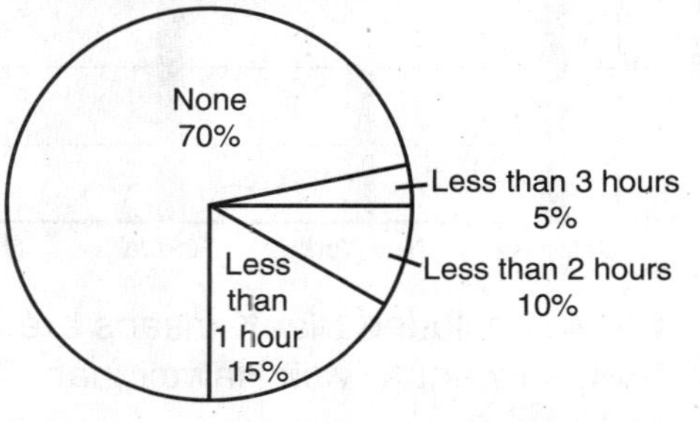

4. Complete the checklist:

a. What population is Lucy studying?

____

b. Is a sample taken from this population? If so, what kind of sample?

____

____

c. Is the survey question unbiased? ______

5. Fill in the blank: Lucy claims that ______ of students do not watch TV.

6. Is Lucy's claim valid? Explain.

____

____

____

____

____

Name ______________________ Date __________ Class __________

Focus on CAHSEE Standards: Intervention

STANDARD 6PS2.5

Sample Problems

Grade 6 • Statistics, Data Analysis, and Probability 2.5

Identify claims based on statistical data and, in simple cases, evaluate the validity of claims.

State whether the conclusion is valid.

1. Leah randomly selected 210 people from Kansas. She asked them which state they would like to visit within the next year.

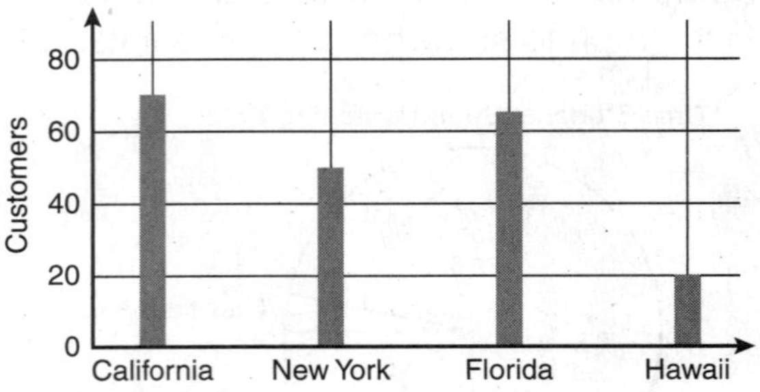

Leah concluded that Kansans are less likely to want to visit Hawaii than California.

What is the population? ______________

____

What kind of sample did Leah take?

____

Does her sample match the population?

______

Is the question biased? ______

Is Leah's graph misleading? ______

Did Leah come to a valid conclusion?

______

The results of two surveys are pictured below. Match each circle graph with its question.

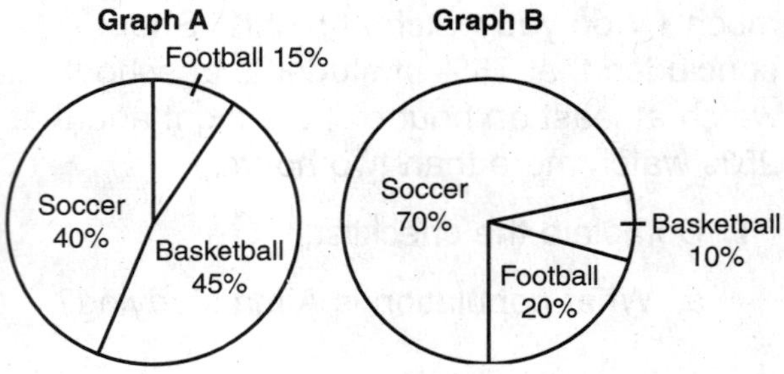

2. "Would you rather play the world's favorite sport, soccer, or would you rather play basketball or football?"

____

Does the question lead you to answer in a certain way? ______

Which of the 2 graphs best fits the answer you could expect from the question?

____

3. "Would you rather play a game like football that requires lots of equipment, or would you rather play a game that only requires a ball, like soccer or basketball?"

____

Does the question lead you to answer in a certain way? ______

Which of the 2 graphs best fits the answer you could expect from the question?

____

Name ______________________ Date ____________ Class ____________

Focus on CAHSEE Standards: Intervention

STANDARD 6PS2.5

Skills Practice

Grade 6 • Statistics, Data Analysis, and Probability 2.5

Identify claims based on statistical data and, in simple cases, evaluate the validity of claims.

Each data display shows the result of a survey of 500 randomly selected sixth-graders.

Survey question: Zipster Sneakers are the best-selling sneakers in the world. What is your favorite sneaker?

1. Is the graph misleading? Explain.

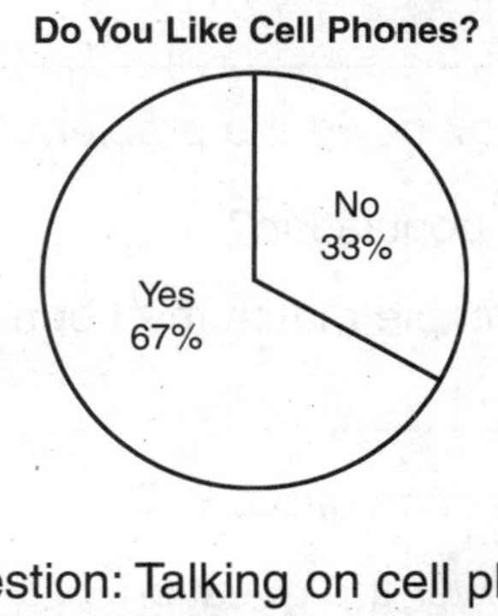

Survey question: Talking on cell phones while driving is a leading cause of car accidents. Do you like cell phones?

2. Is the graph misleading? Explain.

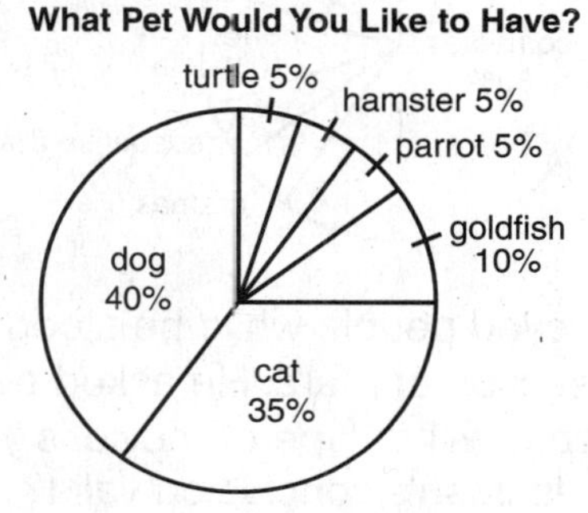

Survey question: What kind of pet would you like to have?

3. Is the graph misleading? Explain.

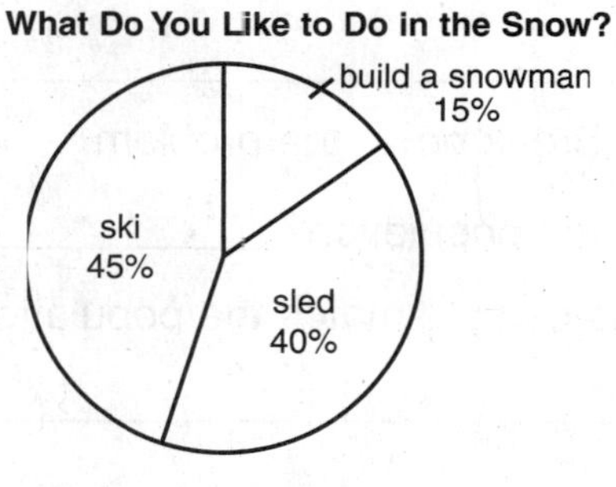

Survey question: What do you like to do in the snow?

4. Is the graph misleading? Explain.

Name ____________________ Date __________ Class __________

STANDARD 6PS2.5

Focus on CAHSEE Standards: Intervention

Problem Solving Practice

Grade 6 • Statistics, Data Analysis, and Probability 2.5

Identify claims based on statistical data and, in simple cases, evaluate the validity of claims.

1. Josh took a survey in Mel's Video Store and then made the following circle graph:

Favorite Types of Movies of Everyone in Mel's Video Store

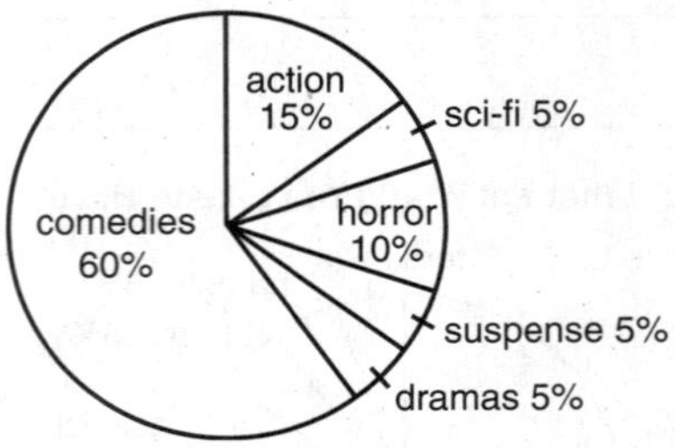

Josh surveyed people while he stood in the comedy section of Mel's. He asked every third person, "What type of movie is your favorite?" Is Josh's conclusion valid?

Step 1: What are you asked to do? ____________________

Step 2: What are the facts?

Where was the survey taken? ____________________

What was Josh's conclusion? ____________________

Step 3: Break down the problem.

What is the population? ____________________

Does the sample match the population?

____

Was his question unbiased? ____________________

Step 4: Answer the question.

Is Josh's conclusion valid? ____________________

Step 5: Check that your answer makes sense.

2. Sam took a survey to find out what cheeses his customers would most like him to put on sale, and then he made this bar graph:

Cheeses My Customers Want on Sale

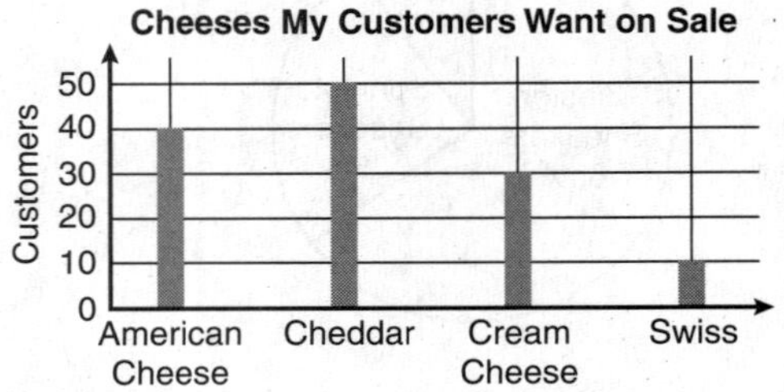

Sam surveyed every tenth person who walked by the store but did not come inside.

Is Sam's data valid?

Step 1: What are you asked to do? ____________________

Step 2: What are the facts?

Where was the survey taken? ____________________

What does the graph show? ____________________

Step 3: Break down the problem.

What is the population? ____________________

Does the sample match the population?

____

Was his question unbiased? ____________________

Step 4: Answer the question.

Is Sam's data and conclusion vaild? ____________________

Step 5: Check that your answer makes sense.

Name ______________________ Date __________ Class __________

Focus on CAHSEE Standards: Intervention

STANDARD 6PS3.1

Alternative Strategies

Grade 6 • Statistics, Data Analysis, and Probability 3.1

Represent all possible outcomes for compound events in an organized way (e.g. tables, grids, tree diagrams) and express the theoretical probability of each outcome.

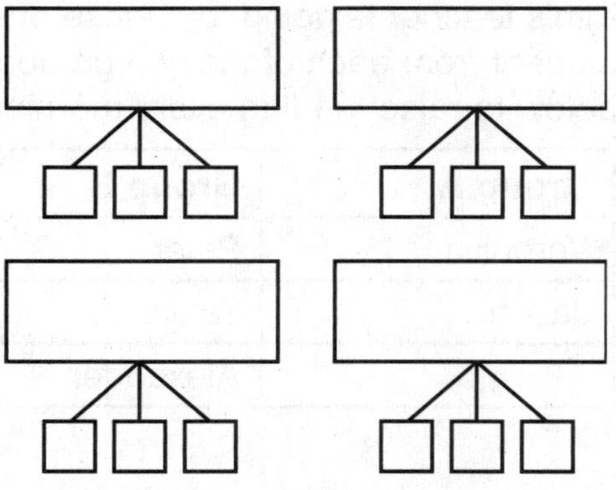

Use the tree diagram.

1. Fill in the tree diagram above to show that red, white, blue, and green T-shirts are available in small, medium, and large.

   a. Write the colors in the rectangles.

   b. Write the sizes in the squares. Use the letters S, M, and L.

2. Use a separate sheet of paper to make another tree diagram. Use it to show that sizes small, medium, large, and extra-large tank tops are available in the colors red, black, and white.

   a. Write the sizes in the rectangles.

   b. Write the colors in the squares. Use the letters R, B, and W.

	A	B	C	D	E
A	AA	AB	AC	AD	AE
B	BA	BB			
C					
D					
E					

Use the chart.

3. Five people are named Alice, Betty, Charlie, Dan, and Eugene. Each person shakes hands with each other person. Make a chart to show all the possible handshakes.

   a. Complete the squares in the chart. Each two-letter pair shows one hand shake.

   b. Cross out the pairs that show a person shaking hands with herself or himself.

   c. Draw a diagonal line through AA, BB, CC, DD, and EE.

   d. The pairs below the diagonal are duplicates. For example, you don't need both AB and BA. Cross out the duplicates below the diagonal.

Name ______________________ Date ____________ Class ____________

STANDARD 6PS3.1

Focus on CAHSEE Standards: Intervention

Sample Problems

Grade 6 • Statistics, Data Analysis, and Probability 3.1

Represent all possible outcomes for compound events in an organized way (e.g. tables, grids, tree diagrams) and express the theoretical probability of each outcome.

1.

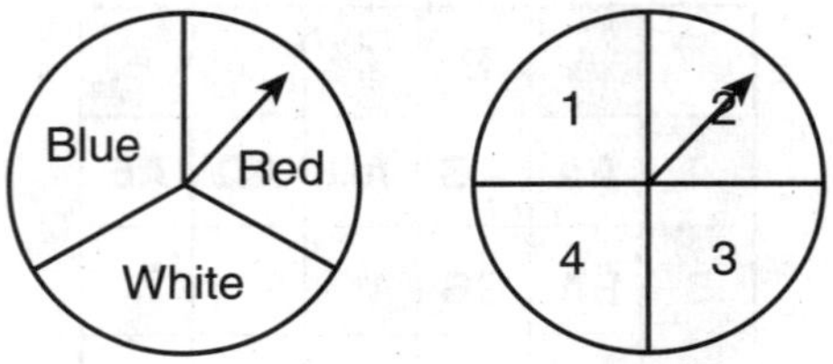

If Amy spins each spinner once, how many possible combinations can she make?

Complete the tree diagram.

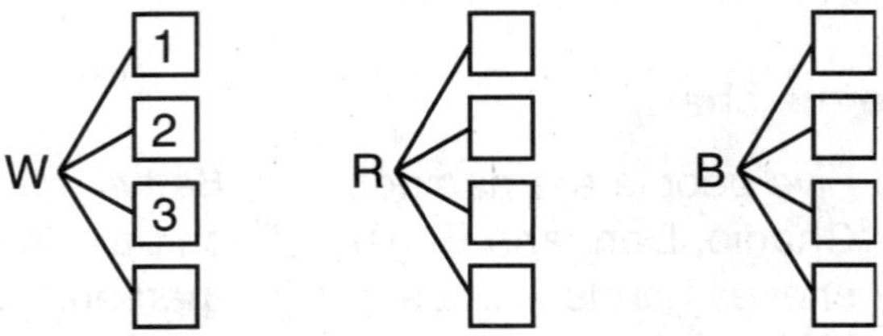

How many different combinations are there?

___

You can also find the number of different combinations by multiplying.

How many different colors are there? ___

How many different numbers are there? ___

3 • 4 = ___

2. Tim's teacher is going to choose one student from each of the two groups below to raise the flag every morning.

Group A	Group B
Veronica	Peter
Jason	Tanya
	Alexander

How many combinations of students does Tim's teacher have to choose from?

To find the number of combinations, multiply the number of students from each group.

Number of students in Group A = 2

Number of students in Group B = 3

___ • ___ = ___

There are ___ combinations from which the teacher can choose.

3. Melinda has 2 pairs of shoes, 3 pairs of pants, and 5 shirts. If Melinda wants to wear a different combination every day, how many days will it take her to wear every combination?

To find the number of combinations, multiply the number of choices of shoes, pants, and shirts.

Number of pairs of shoes = ___

Number of pairs of pants = ___

Number of shirts = ___

___ • ___ • ___ = ___

It will take Melinda ___ days to wear every possible combination.

Name ______________________________ Date ____________ Class ____________

STANDARD 6PS3.1

Focus on CAHSEE Standards: Intervention

Skills Practice

Grade 6 • Statistics, Data Analysis, and Probability 3.1

Represent all possible outcomes for compound events in an organized way (e.g. tables, grids, tree diagrams) and express the theoretical probability of each outcome.

1. Helena has been asked to choose what food to bring to the school picnic. She needs to choose one vegetable, one main dish, and one dessert from the lists below.

Vegetable	Main Dish	Dessert
Corn	Pasta	Melon
Beans	Chicken	Oranges

How many combinations does Helena have to choose from? ____

2. Terry has 2 pairs of shoes, 2 pairs of pants, and 3 shirts. How many different combinations can he wear? ____

3. Anna is eating soup and salad for lunch. She can choose from chicken noodle, tomato, or squash soup. She can choose a garden, spinach, or Greek salad. How many different lunches can Anna eat?

____

4. In the school election, Alana, Lisa, and Tina are running for class president. Fred and Kate are running for vice president. Create a table that shows the different combinations of election results.

President	Vice President

5. The school store is selling T-shirts in red, blue, yellow, and green. There are 3 sizes to choose from: small, medium, and large. Create a tree diagram to show how many different combinations of T-shirts there are.

Write each fraction in simplest form.

6. $\frac{3}{12}$ ______ **7.** $\frac{3}{9}$ ______

8. $\frac{4}{8}$ ______ **9.** $\frac{12}{72}$ ______

Change the following fractions into decimals. Round to the nearest hundredth.

10. $\frac{1}{2}$ ______ **11.** $\frac{1}{3}$ ______

12. $\frac{4}{5}$ ______ **13.** $\frac{3}{8}$ ______

Change the following decimals into percents.

14. 0.28 ______ **15.** 0.13 ______

16. 0.33 ______ **17.** 0.06 ______

Name ______________________ Date ____________ Class ____________

STANDARD **6PS3.1**

Focus on CAHSEE Standards: Intervention

Problem Solving Practice

Grade 6 • Statistics, Data Analysis, and Probability 3.1

Represent all possible outcomes for compound events in an organized way (e.g. tables, grids, tree diagrams) and express the theoretical probability of each outcome.

1. Andres spins a spinner that is labeled 1–8. If Andres spins the spinner once, what is the probability that he will spin a 6? Express the probability as a decimal. Round to the nearest hundredth. Express the probability as a percent.

Step 1: What are you asked to find?

____

____

Step 2: What are the facts?
The spinner is labeled 1–8.
He will spin the spinner once.

Step 3: Break down the problem.
What are the possible outcomes?

____

What is the number of possible outcomes? ___

What is the number of favorable outcomes? ___

Step 4: Solve the problem.
Use this formula to find the probability as a fraction:

$$P(6) = \frac{\text{number of favorable outcomes}}{\text{number of possible outcomes}}$$

$$P(6) = \frac{\square}{\square}$$

Express the probability as a decimal.

___ ÷ ___ = _______ = _______

Express the probability as a percent.

_______ = ___%

Step 5: Check that your answer makes sense. Estimate the probability of spinning a 6 by shading $\frac{1}{8}$ of a diagram of a spinner.

2. Nina has a jar containing 4 blue marbles and 6 green marbles. What is the probability that she will pick a blue marble without looking? Express your answer as a fraction, decimal, and percent.

Step 1: What are you asked to find?

____

____

Step 2: What are the facts?

Color	Number of Marbles
Blue	4
Green	6

Step 3: Break down the problem.

What is the number of possible outcomes? ___

What is the number of favorable outcomes? ___

Step 4: Solve the problem.

$$P(\text{blue}) = \frac{\square}{\square} = \frac{\square}{\square}$$

Express the probability as a decimal and a percent.

___ ÷ ___ = _______ = ___%

Step 5: Check your answer makes sense.
Find the probability of picking a green marble.

$$P(\text{green}) = \frac{\square}{\square} = \frac{\square}{\square} = \_\_\_\_\_\_ = \_\_\_\%$$

P(blue) + P(green) should equal 100%.

Does it? _______

Name ______________________ Date __________ Class __________

STANDARD 6PS3.3

Focus on CAHSEE Standards: Intervention

Alternative Strategies

Grade 6 • Statistics, Data Analysis, and Probability 3.3

Represent probabilities as ratios, proportions, decimals between 0 and 1, and percentages between 0 and 100 and verify that the probabilities computed are reasonable; know that if *P* is the probability of an event 1–*P* is the probability of an event not occurring.

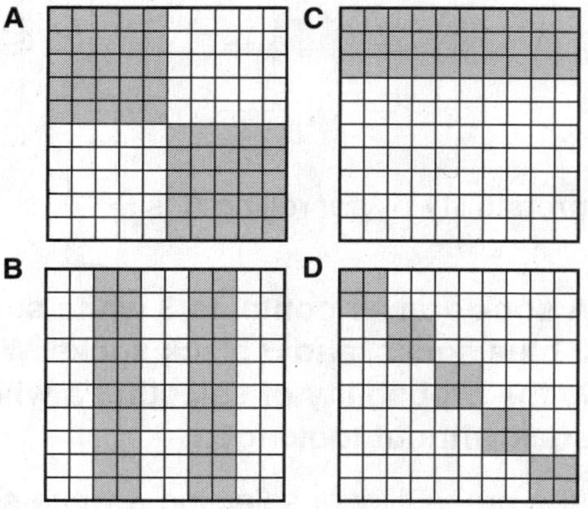

Use the grids.

1. If an object is dropped on to square A, what is the probability of it landing on a shaded part?

   a. Use 100 as the denominator. ______

   b. Write in simplest form. ______

2. If an object is dropped on to square B, what is the probability of it landing on a shaded part?

   a. Use 100 as the denominator. ______

   b. Write in simplest form. ______

3. Write the probability in simplest form of an object landing on the shaded part.

   a. square C ______

   b. square D ______

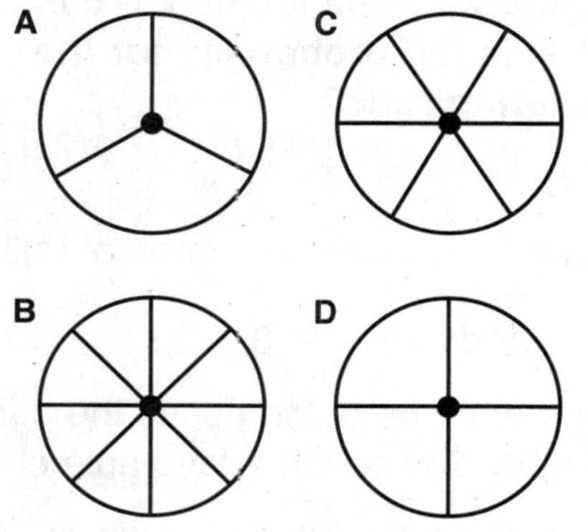

Use the spinners.

4. Shade one part of spinner A. Then find the probability of a spinner landing on

   a. the shaded part ______

   b. a white part ______

5. Shade two parts of spinner B. Write the probability in simplest form of landing on

   a. the shaded part ______

   b. a white part ______

6. Shade part of spinner C so that the probability of landing on the shaded part is $\frac{1}{3}$.

7. Shade part of spinner D so that the probability of NOT landing on the shaded part is $\frac{1}{4}$.

Name ______________________________ Date __________ Class __________

Focus on CAHSEE Standards: Intervention

STANDARD 6PS3.3 *Sample Problems*

Grade 6 • Statistics, Data Analysis, and Probability 3.3

Represent probabilities as ratios, proportions, decimals between 0 and 1, and percentages between 0 and 100 and verify that the probabilities computed are reasonable; know that if P is the probability of an event $1-P$ is the probability of an event not occurring.

1. A number cube labeled 1–6 is rolled. What is the probability for the following rolls?

a. *P*(2)

There are 6 possible outcomes of rolling the number cube: 1, 2, 3, ___, ___, and ___.

Only one of these outcomes is the number 2. The number 2 is called a favorable outcome.

$P(2) = \frac{\text{number of favorable outcomes}}{\text{number of possible outcomes}}$

$P(2) = \frac{\square}{\square}$

b. *P*(even)

What are the possible even numbers?

____

How many favorable outcomes are there? ___

How many possible outcomes are there? ___

$P(\text{even}) = \frac{\text{number of favorable outcomes}}{\text{number of possible outcomes}}$

$P(\text{even}) = \frac{\square}{\square} = \frac{\square}{\square}$

c. *P*(not a 5)

Find the probability of rolling 5.

$P(5) = \frac{\square}{\square}$

In order to find the probability of *not* rolling 5, subtract the probability of rolling 5 from 1.

$P(\text{not } 5) = 1 - P(5)$

$P(\text{not } 5) = 1 - \frac{\square}{\square} = \frac{\square}{\square} - \frac{\square}{\square} = \frac{\square}{\square}$

The probability of *not* rolling 5 is $\frac{\square}{\square}$.

2. A sock drawer contains 3 white socks, 2 blue socks, and 5 black socks. What is the probability of selecting a white sock without looking?

Find the probability of selecting a white sock from the drawer.

How many favorable outcomes are there? ___

How many possible outcomes are there? ___

To write the probability of choosing a white sock from the drawer, use the following formula:

$P(\text{white}) = \frac{\text{number of favorable outcomes}}{\text{number of possible outcomes}}$

$P(\text{white}) = \frac{\square}{\square}$

Show the probability as a ratio.

P(white) = ___ to ___

Show the probability as a decimal between 0 and 100.

$P(\text{white}) = \frac{\square}{\square} =$ ______

Show the probability as a percent.

P(white) = ______ = ___%

Name ______________________________ Date ____________ Class ____________

STANDARD 6PS3.3

Focus on CAHSEE Standards: Intervention

Skills Practice

Grade 6 • Statistics, Data Analysis, and Probability 3.3

Represent probabilities as ratios, proportions, decimals between 0 and 1, and percentages between 0 and 100 and verify that the probabilities computed are reasonable; know that if *P* is the probability of an event 1–*P* is the probability of an event not occurring.

A number cube labeled 1–6 is rolled. Find each of the following probabilities. Write the probability as a fraction.

1. *P*(odd) ______
2. *P*(1 or 3) ______
3. *P*(not 1 or 5) ______
4. *P*(not odd) ______
5. *P*(6) ______
6. *P*(2, 3, or 4) ______

A spinner labeled 1–9 is spun. Find each of the following probabilities. Write the probability as a fraction.

7. *P*(1, 2, or 9) ______
8. *P*(even) ______
9. *P*(2) ______
10. *P*(not 3) ______
11. *P*(7 or 8) ______
12. *P*(not 6, 7, or 8) ______

A spinner labeled 1–15 is spun. Find each of the following probabilities. Write the probability as a decimal. Round to the nearest hundredth.

13. *P*(odd) ______
14. *P*(1, 2, 3, or 4) ______
15. *P*(6 or 10) ______

A spinner labeled 1–15 is spun. Find each of the following probabilities. Write the probability as a decimal.

16. *P*(even) ______
17. *P*(not 3 or 4) ______
18. *P*(15) ______

A spinner labeled 1–20 is spun. Find each of the following probabilities. Write the probability as a percent.

19. *P*(2 or 20) ______
20. *P*(18, 19, or 20) ______
21. *P*(3) ______
22. *P*(not 1 or 2) ______

Name ______________________ Date __________ Class __________

STANDARD 6PS3.3

Focus on CAHSEE Standards: Intervention

Problem Solving Practice

Grade 6 • Statistics, Data Analysis, and Probability 3.3

Represent probabilities as ratios, proportions, decimals between 0 and 1, and percentages between 0 and 100 and verify that the probabilities computed are reasonable; know that if *P* is the probability of an event 1–*P* is the probability of an event not occurring.

1. John has a bag containing marbles: 5 orange, 27 red, and 18 blue. What is the probability that, without looking, John will pick a blue marble?

Step 1: What are you asked to find?

____

____

Step 2: What are the facts?

Color	Number of Marbles
Orange	5
Red	____
Blue	____

Step 3: Break down the problem.
What is the number of favorable outcomes?

___ Total possible outcomes? ___

Step 4: Solve the problem.
Use this formula to find the probability:

$$P(\text{blue}) = \frac{\text{number of favorable outcomes}}{\text{number of possible outcomes}}$$

$$P(\text{blue}) = \frac{\square}{\square} = \frac{\square}{\square}$$

Step 5: Check that your answer makes sense.
Find the probability of *not* picking blue.

$$1 - \frac{\square}{\square} = \frac{\square}{\square} - \frac{\square}{\square} = \frac{\square}{\square}$$

The probability should be the same as picking an orange or red marble.

2. Natasha has 15 T-shirts in a box: 7 yellow, 3 green, 3 red, and 2 black. What is the probability that, without looking, Natasha will *not* pick a green T-shirt?

Step 1: What are you asked to find?

____

____

Step 2: What are the facts?

Color	Number of T-Shirts
Yellow	____
Green	____
Red	____
Black	____

Step 3: Break down the problem.
What is the number of favorable outcomes?

___ Total possible outcomes? ___

Step 4: Solve the problem.

$$P(\text{green}) = \frac{\square}{\square} = \frac{\square}{\square}$$

$$P(\text{not green}) = 1 - P(\text{green}) = 1 - \frac{\square}{\square}$$

$$P(\text{not green}) = \frac{\square}{\square} - \frac{\square}{\square} = \frac{\square}{\square}$$

Step 5: Does your answer make sense?
Find the probability of picking a yellow, red, or black T-shirt.

$$P(\text{yellow, red, or black}) = \frac{\square}{\square} = \frac{\square}{\square}$$

The probability should be the same as *not* picking a green T-shirt.

Name ______________________________ Date ___________ Class ___________

STANDARD 6PS3.5 Focus on CAHSEE Standards: Intervention

Alternative Strategies

Grade 6 • Statistics, Data Analysis, and Probability 3.5

Understand the difference between independent and dependent events.

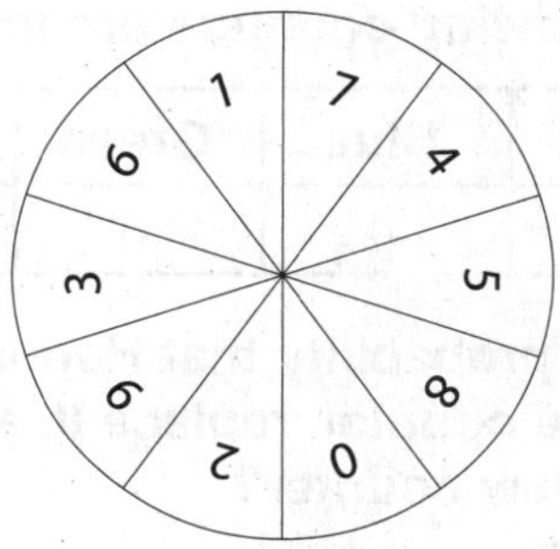

Use the spinner.

1a. How many different outcomes are possible on this spinner? ___

1b. How many numbers on this spinner are greater than 5? ___

1c. Find the probability of spinning a number greater than 5. Simplify the fraction. ___

2a. How many even numbers are on the spinner? How many odd numbers?

____

2b. Find the probability of spinning an even number. Find the probability of spinning an odd number.

____

2c. Find the probability of first spinning an even number and then spinning an odd number on two spins. ___

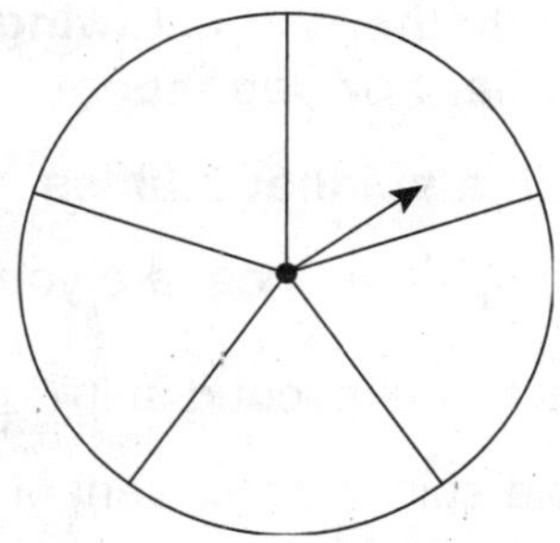

Use the blank spinner.

3a. Write the first 5 letters of the alphabet on the spinner.

3b. Find the probability of spinning a vowel and then a consonant on two spins. ___

4a. Find the probability of flipping a coin and it landing heads up. ___

4b. Find the probability of flipping a coin and it landing tails up and then spinning a C on the spinner. ___

5a. Jake is playing a game. He flips a coin. If it lands heads up, he flips the coin again. If the first flip results in tails, he spins the spinner. On another sheet of paper, make a tree diagram to show all the possible outcomes.

____

5b. Jake has flipped the coin. What is the probability that the next thing he does is spin an E? ___

Name ______________________ Date __________ Class __________

STANDARD 6PS3.5

Focus on CAHSEE Standards: Intervention

Sample Problems

Grade 6 • Statistics, Data Analysis, and Probability 3.5

Understand the difference between independent and dependent events.

Determine whether the following events are independent or dependent.

1. You spin a spinner 3 times.

If you spin a spinner once, are you affecting the outcome of the second spin? __________

Is the second spin a dependent or independent event? __________

2. You select a marble from a can and do not replace it before selecting another marble.

If you select a marble and do not replace it, are you affecting the outcome of the second selection? __________

Is the second selection a dependent or independent event? __________

3. You select two counters in a row from a box without replacing them.

If you select a counter and don't replace it, are you affecting the number of possible outcomes for the second selection? __________

Is the second selection a dependent or independent event? __________

4. The following counters are in a box.

Yellow	Blue	Green	Red
2	4	3	1

What is the probability that Hannah will select a blue counter, replace it, and then select a yellow counter?

Find the probability of the first selection.

The number of favorable outcomes for selecting a blue counter is 4.

The number of possible outcomes is 10.

$$P(\text{blue}) = \frac{\text{number of favorable outcomes}}{\text{number of possible outcomes}}$$

$$P(\text{blue}) = \frac{4}{10} = \frac{2}{5}$$

The blue counter is replaced for the second selection. The events are __________.

Find the probability of the second selection.

The number of favorable outcomes for selecting a yellow counter is ___.

The number of possible outcomes for the second selection is ___.

$$P(\text{yellow}) = \frac{\text{number of favorable outcomes}}{\text{number of possible outcomes}}$$

$$P(\text{yellow}) = \frac{\square}{\square} = \frac{\square}{\square}$$

Find the probability of both events occurring.

$$P(\text{blue and yellow}) = \frac{2}{5} \bullet \frac{\square}{\square} = \frac{\square}{\square}$$

The probability of Hannah selecting a blue counter and then a yellow counter is ___.

Name ______________________ Date ____________ Class ____________

STANDARD 6PS3.5

Focus on CAHSEE Standards: Intervention

Skills Practice

Grade 6 • Statistics, Data Analysis, and Probability 3.5

Understand the difference between independent and dependent events.

Determine whether the following events are independent or dependent.

1. You flip a coin five times.

   ____

2. You roll a cube and then spin a spinner.

   ____

3. You select a name from a hat, do not replace it, and then select another name.

   ____

4. You choose a flash card randomly from a pile, replace it, and then choose another card.

   ____

5. You take two tennis balls in a row from a gym bag without putting any of them back.

   ____

6. You select a marble from a jar and then select a second marble without replacing the first.

   ____

7. You pick an apple out of a crate, put it back, and pick a second apple.

   ____

8. You roll a 6-sided cube and then flip a coin 4 times.

   ____

9. You choose a bean from a cup, hand it to a friend, and then choose another bean from the cup.

   ____

10. You select a raffle ticket from a bag. Then you put the raffle ticket back in the bag before selecting a second one.

    ____

11. You spin a spinner that is labeled 1–8, then you spin another spinner that is labeled 1–7.

    ____

12. You select one counter from a box, replace it, and then select another counter.

    ____

13. You choose a card from a deck and then choose another card without replacing the first.

    ____

14. You randomly select a button from a can, but you do not replace it before selecting another.

    ____

15. You toss a coin, spin a spinner two times, and then toss the coin a second time.

    ____

16. You randomly pick a counter from a bag, replace it, and then randomly pick another counter.

    ____

Name ______________________ Date __________ Class __________

STANDARD 6PS3.5

Focus on CAHSEE Standards: Intervention

Problem Solving Practice

Grade 6 • Statistics, Data Analysis, and Probability 3.5

Understand the difference between independent and dependent events.

1. The following tennis balls are in a gym bag: 7 green, 5 orange, and 3 white.

What is the probability of Jake selecting an orange tennis ball, and then, without replacing the orange ball, selecting a green tennis ball?

Step 1: What are you asked to find?

____

____

Step 2: What are the facts?

Color	Number of Balls
Green	7
Orange	___
________	___
	Total = ___

Step 3: Break down the problem.

For the first selection, there are ___ possible outcomes and ___ favorable outcomes.

For the second, there are ___ possible outcomes and ___ favorable outcomes.

Step 4: Solve the problem.
Find the probability of the two selections:

$P(\text{orange}) = \frac{\square}{\square} = \frac{\square}{\square}$

$P(\text{green}) = \frac{\square}{\square} = \frac{\square}{\square}$

Find the probability of the two events.

$P(\text{green after orange}) = \frac{\square}{\square} \cdot \frac{\square}{\square} = \frac{\square}{\square}$

Step 5: Check that your answer makes sense.

2. The following marbles are in a jar: 4 red, 6 blue, and 10 green.

What is the probability of Marie selecting a green marble, and then, upon replacing the green marble, selecting a blue marble?

Step 1: What are you asked to find?

____

____

Step 2: What are the facts?

Color	Number of Marbles
Red	___
Blue	___
Green	___
	Total = ___

Step 3: Break down the problem.

For the first selection, there are ___ possible outcomes and ___ favorable outcomes.

For the second, there are ___ possible outcomes and ___ favorable outcomes.

Step 4: Solve the problem.
Find the probability of the two selections:

$P(\text{green}) = \frac{\square}{\square} = \frac{\square}{\square}$

$P(\text{blue}) = \frac{\square}{\square} = \frac{\square}{\square}$

Find the probability of the two events.

$P(\text{green and blue}) = \frac{\square}{\square} \cdot \frac{\square}{\square} = \frac{\square}{\square}$

Step 5: Check that your answer makes sense.

Name ________________________________ Date __________ Class __________

STANDARD **7NS1.1**

Focus on CAHSEE Standards: Intervention

Sample Problems

Grade 1 • Number Sense 1.1

Read, write, and compare rational numbers in scientific notation, compare rational numbers.

1. Write the number 324 in scientific notation.

Write 324 as a decimal: ______

Move the decimal point to the left so that 324 becomes a number that is at least 1 but less than 10. What is that number? ______

How many places did you move the decimal point? ____

Write an equation to illustrate how many places you moved the decimal point:

324 = ____.____ × 10 × ____

How do you write 100 using powers of 10? ____

Write 324 using scientific notation:

____.____ × 10__

2. Write the number 768,390 in scientific notation.

Write 768,390 as a decimal: __________

Move the decimal so that 768,390 becomes at least 1 but less than 10.

What is that number? __________

How many places did you move the decimal point? ____

Write an equation to illustrate how many places you moved the decimal point:

768390 = ____________________

____

Write 768,390 using scientific notation:

____

3. Write the number 0.034 in scientific notation.

Move the decimal to the right so that 0.034 becomes at least 1 but less than 10. What is that number? ______

How many places did you move? ____

Moving the decimal 2 places to the right is equivalent to dividing by the number ______.

Write an equation to illustrate how many places you moved the decimal point:

0.034 = ______________

Dividing by 100 is the same as multiplying by 10^{-2} since the decimal place moves right.

Write 0.034 using scientific notation:

____

Name ______________________ Date __________ Class __________

STANDARD **7NS1.1**

Focus on CAHSEE Standards: Intervention

Skills Practice

Grade 7 • Number Sense 1.1

Read, write, and compare rational numbers in scientific notation (positive and negative powers of 10), compare rational numbers.

Write each number in scientific notation.

1. 859 __________
2. 18,000 __________
3. 0.076 __________
4. 0.0057 __________
5. 755,000 __________
6. 5,000,000 __________
7. The population of the New York City is approximately 8 million. Write the population in scientific notation.

   ____
8. The population of California is approximately 29 million. Write the population in scientific notation.

   ____
9. Earth is approximately 93 million miles from the sun. Write the distance in scientific notation.

   ____
10. A small monarch butterfly weighs approximately 0.0095 ounces. Write the weight in scientific notation.

    ____

Write each number in standard form.

11. 5.31×10^3 __________
12. 7.9×10^4 __________
13. 9.77×10^5 __________
14. 4.25×10^5 __________
15. 3.96×10^4 __________
16. 2.22×10^6 __________
17. Mel's car weighs one ton, or 3.2×10^4 ounces. Write the weight in standard form.

    ____
18. 3.3×10^{-4} __________
19. 8.2×10^{-5} __________
20. 5.55×10^{-3} __________
21. 4.82×10^{-2} __________
22. 6.1×10^{-3} __________
23. 2.5×10^{-6} __________
24. The diameter of a bolt is 1.25×10^{-1} cm. Write the diameter in standard form.

    ____

Name ______________________ Date __________ Class __________

STANDARD 7NS1.2

Focus on CAHSEE Standards: Intervention

Alternative Strategies

Grade 7 • Number Sense 1.2

Add, subtract, multiply, and divide rational numbers (integers, fractions, and terminating decimals) and take positive rational numbers to whole-number powers.

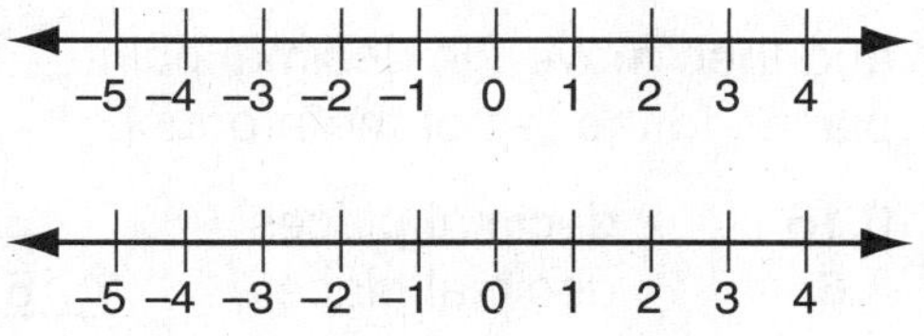

Use the number lines.

1a. Represent 3 on the number line using an arrow starting at 0 and going to 3.

1b. Represent 3 + (–4) on the number line with an arrow starting at 3 and going left 4 units.

1c. Where does the arrow end? ____

1d. What is the sum 3 + (–4)? ____

2a. Represent 3 on the number line using an arrow.

2b. Represent 3 – (–1) on the number line.

2c. What is the difference 3 – (–1)?

____

3. Use a number line to show and find –6 ÷ 3.

____

____

____

____

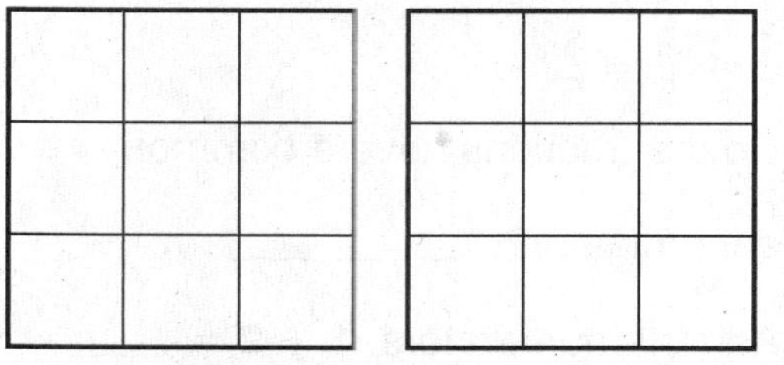

Use the squares.

4a. Fill the nine small squares of one 3-by-3 square with two "$1" and seven "debts of $1."

4b. What sum does this give you?

____

4c. How much are you in debt?

____

5a. Fill the nine small squares of the other 3-by-3 square with two "$1" and seven "debts of $1."

5b. How much debt does this show?

____

5c. What product shows how much you are in debt now?

____

Name ______________________ Date ____________ Class ____________

STANDARD 7NS1.2

Focus on CAHSEE Standards: Intervention

Sample Problems

Grade 7 • Number Sense 1.2

Add, subtract, multiply, and divide rational numbers (integers, fractions, and terminating decimals) and take positive rational numbers to whole-number powers.

Add.

1. $\frac{1}{5} + \frac{2}{5} = \frac{3}{5}$

Do the fractions have a common denominator? __________

Add the numerators: $1 + 2 =$ ____

Write the sum of the numerators over the common denominator. ____

Subtract.

2. $(-3) - (-4)$

(-3) and (-4) are __________ integers.

Use the additive inverse: $-3 +$ ____

What is $(-3) - (-4)$? ____

Multiply.

3. $2\frac{1}{7} \times 2\frac{1}{3}$

Rewrite as improper fractions: ________

Multiply the numerators: ________ = ____

Multiply the denominators:

________ = ____

Write the product of the numerators over the product of the denominators, and simplify: $2\frac{1}{7} \times 2\frac{1}{3} =$ ____ = ____

4. $0.15 \cdot 1.6$

There are ____ decimal places.

Multiply as if there is no decimal point, and then move the decimal point ____ places to the left of the product.

0.15	2 decimal places	15
1.6	1 decimal place	× 16
		+ ____

Move decimal 3 places to the left: ____

5. $10^2 \cdot 10^5$

Add the exponents: __________

Rewrite the sum of the exponents so the number is an exponent: ____

Divide.

6. $\frac{7}{9} \div 2$

Rewrite as improper fractions: $\frac{7}{9} \div$ ____

Multiply by the reciprocal: $\frac{7}{9} \cdot$ ____

Multiply the numerators: $7 \cdot$ ____ = ____

Multiply the denominators:

$9 \cdot$ ____ = ____

Write the product of the numerators over the product of the denominators and reduce: __________________

____

Name ______________________ Date ____________ Class ____________

STANDARD 7NS1.2

Focus on CAHSEE Standards: Intervention

Skills Practice

Grade 7 • Number Sense 1.2

Add, subtract, multiply, and divide rational numbers (integers, fractions, and terminating decimals) and take positive rational numbers to whole-number powers.

Add.

1. $8 + (-3) =$ ____
2. $(-3) + (-4) =$ ____
3. $(-5) + (-6) =$ ____
4. $\frac{2}{11} + \frac{7}{11} =$ ____
5. $1\frac{3}{7} + 3\frac{2}{7} =$ ____
6. $3.83 + 22 =$ ____

Subtract.

7. $(-3) - (-4) =$ ____
8. $(-5) - (-6) =$ ____
9. $3 - 4 =$ ____
10. $\frac{2}{5} - \frac{1}{5} =$ ____
11. $4\frac{1}{6} - 2\frac{5}{6} =$ ____
12. $3.83 - 22 =$ ____
13. $1.4505 - (-1.45)$ ____

Multiply.

14. $\frac{3}{5} \times \frac{3}{8} =$ ____
15. $\frac{2}{7} \times \frac{4}{9} =$ ____
16. $10^6 \times 10^2 =$ ____
17. $2\frac{1}{5} \times 2\frac{2}{3} =$ ____
18. $3.83 \cdot 22 =$ ____
19. $111 \cdot 1.11 =$ ____
20. $106 \cdot 1.02 =$ ____
21. $123 \cdot 3.21 =$ ____
22. $152 \cdot 5.31 =$ ____

Divide.

23. $\frac{2}{5} \div \frac{1}{5} =$ ____
24. $1\frac{3}{5} \div 3\frac{1}{3} =$ ____
25. $2\frac{4}{5} \div \frac{1}{3} =$ ____
26. $1\frac{5}{6} \div \frac{1}{5} =$ ____
27. $40 \div 1.5 =$ ____
28. $1.5 \div 40 =$ ____
29. $10^6 \div 10^2 =$ ____

Name ______________________________ Date ____________ Class ____________

STANDARD 7NS1.2

Focus on CAHSEE Standards: Intervention

Problem Solving Practice

Grade 7 • Number Sense 1.2

Add, subtract, multiply, and divide rational numbers (integers, fractions, and terminating decimals) and take positive rational numbers to whole-number powers.

Erica started her trip to her parents' house with $\frac{3}{4}$ of a tank of gasoline. When she arrived, her gas tank was $\frac{1}{4}$ full. What part of a full tank of gasoline did she use on her trip?

Step 1: What are you asked to find?

____

____

Step 2: What are the facts?

start with gas tank $\frac{3}{\square}$ full

arrive with gas tank $\frac{\square}{4}$ full

Step 3: Break down the problem.
(part of tank with gasoline at start) −
(part of tank full arriving at parents' house) =

____

Step 4: Solve the problem.

____ $- \frac{1}{4} =$ ______________

The part of a full tank of gasoline that she used is ____ tank.

Step 5: Check your answer by using addition. Add the amount of gas Erica used to the amount she had left. The answer should equal the amount she had to start with.

____ $+ \frac{1}{4} = \frac{3}{4}$

Doug spent $2.50 at a car wash to vacuum the interior of Erica's car before her trip. If 50 cents operates the vacuum for 1 minute, how long did Doug vacuum Erica's car?

Step 1: What are you asked to find?

____

____

Step 2: What are the facts?

He spent ____ cents per minute

He spent a total of ______

Step 3: Break down the problem.
Divide the total amount of money he spent by the amount it cost per minute:

______ ÷ ______ = number of minutes he vacuumed

Step 4: Solve the problem.
(Total spent) ÷ (Cost per minute) =
Number of minutes

______ ÷ ______ = _____

____

Doug spent ____ minutes vacuuming the car.

Step 5: Check your answer by using multiplication. Multiply the amount of time he spent vacuuming ____ by the amount of money it costs per minute ______. The answer should equal how much he spent ______.

Name ______________________ Date __________ Class __________

STANDARD 7NS1.3

Focus on CAHSEE Standards: Intervention

Sample Problems

Grade 7 • Number Sense 1.3

Convert fractions to decimals and percents and use these representations in estimations, computations, and applications.

1. Write $\frac{1}{2}$ as a decimal.

First write an equivalent fraction with a denominator that is a multiple of 10.

$\frac{1}{2} \times \frac{5}{\square} = \frac{5}{\square}$

Then write the equivalent fraction as a decimal. $\frac{5}{10} =$ ____

2. Write $\frac{3}{4}$ as a decimal.

Is 4 a factor of 10? ____ of 100? ____

$4 \times$ ____ $= 100$

Write $\frac{3}{4}$ as an equivalent fraction whose denominator is 100. Then write the equivalent fraction as a decimal.

$\frac{3}{4} \times \frac{25}{\square} = \frac{75}{\square} = 0.75$

3. Write $\frac{5}{8}$ as an equivalent fraction whose denominator is 1,000. Then write it as a decimal.

$8 \times$ ____ $= 1000$

$\frac{5}{8} \times \frac{125}{\square} = \frac{\square}{1000} =$ ________

A fraction can be written as a decimal by using division. $\frac{3}{5}$ means $3 \div 5$.

```
  0.6
5)3.0
  3.0
```

The fraction $\frac{3}{5}$ can be written as the decimal 0.6. The division ends because the remainder is zero. A decimal like 0.6 is called a terminating decimal.

4. Use division to find the decimal for $\frac{7}{25}$.

```
25)7.00
   5 0
   200
   200
   000
```

Some fractions cannot be written as a terminating decimal. Express $\frac{1}{3}$ as a decimal using division.

```
  0.333...
3)1.0000
    9
    10
     9
    10
     9
```

Decimals like 0.333333... are called repeating decimals because the pattern in the digits repeats forever. The bar notation $0.\overline{3}$ shows that the 3 repeats forever.

What digit repeats as the division is repeated? ____

5. Since 1982, the alloy used to make the US penny is 0.975 zinc and 0.025 copper. Write each decimal as a percent.

Decimals can be written as a percent by moving the decimal point two places to the right.

0.975 = ______%

0.025 = ______%

Name ______________________ Date ___________ Class ___________

STANDARD 7NS1.3

Focus on CAHSEE Standards: Intervention

Skills Practice

Grade 7 • Number Sense 1.3

Convert fractions to decimals and percents and use these representations in estimations, computations, and applications.

For 1–6, write each fraction as a decimal.

1. $\frac{1}{4}$

2. $\frac{2}{5}$

3. $\frac{7}{20}$

4. $\frac{7}{10}$

5. $\frac{7}{8}$

6. $\frac{6}{25}$

For 7–12, use division to write each fraction as a decimal.

7. $\frac{4}{5}$

8. $\frac{8}{9}$

9. $\frac{1}{6}$

10. $\frac{3}{11}$

11. $\frac{5}{12}$

12. $\frac{7}{11}$

13. In 2006, the California State sales tax rate was 7.25%. What would the sales tax be if you purchased a shirt for $28.00?

14. Approximately $\frac{2}{3}$ of Ted's CD collection is hip-hop music. If he owns 40 CDs, estimate the number of CDs that are hip-hop.

15. Ben's mother wants him to put 4 shoe boxes full of baseball cards in storage. If his collection consists of 12 boxes, about what percentage of his collection must he store?

16. Heather is clearing out her closet and wants to give last year's summer clothes to charity. If she has 16 t-shirts and gives 12 away, what percentage of her t-shirts does she keep?

Name ________________________ Date ____________ Class ____________

STANDARD **7NS1.6**

Focus on CAHSEE Standards: Intervention

Sample Problems

Grade 7 • Number Sense 1.6

Calculate the percentage of increases and decreases of quantity.

Find the percent of increase.
Round to the nearest percent.

1. 48 increased to 86

What is the original amount? ____

What is the formula for percent of change?

percent of change

= ____________________

Find the amount of change or difference.

86 − 48 = ____

Write the percent of change formula using the amounts.

percent of change $= \frac{\square}{\square}$

Write the answer as a decimal and a percent. Round to the nearest percent.

$\frac{38}{48} \approx$ ______ $\approx$ ____%

The percent of increase from 48 to 86

is ____%.

Find the percent of decrease.
Round to the nearest percent.

2. 87 decreased to 24

What is the original amount? ____

What is the formula for percent of change?

percent of change

= ____________________

Find the amount of change or difference.

87 − 24 = ____

Write the percent of change formula using the amounts.

percent of change $= \frac{\square}{\square}$

Write the answer as a decimal and a percent. Round to the nearest percent.

$\frac{63}{87} \approx$ ______ $\approx$ ____%

The percent of decrease from 87 to 24

is ____%.

Name ______________________________ Date ____________ Class ____________

STANDARD 7NS1.6

Focus on CAHSEE Standards: Intervention

Skills Practice

Grade 7 • Number Sense 1.6

Calculate the percentage of increases and decreases of quantity.

Find the percent of increase.
Round to the nearest percent.

1. 23 increased to 46.

______%

2. 22 increased to 88.

______%

3. 35 increased to 47.

______%

4. 48 increased to 104.

______%

5. 96 increased to 98.

______%

6. 52 increased to 68.

______%

Find the percent of decrease.
Round to the nearest percent.

7. 25 decreased to 20.

______%

8. 78 decreased to 52.

______%

9. 125 decreased to 20.

______%

10. 75 decreased to 72.

______%

11. 85 decreased to 79.

______%

12. 105 decreased to 57.

______%

Name ______________________ Date ____________ Class ____________

STANDARD 7NS1.7

Focus on CAHSEE Standards: Intervention

Alternative Strategies

Grade 7 • Number Sense 1.7

Solve problems that involve discounts, markups, commissions, and profit and compute simple and compound interest.

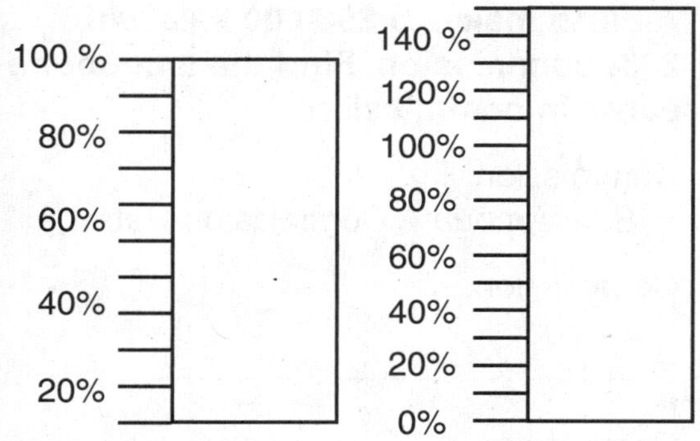

Use the percent of a number diagrams.

1a. Lightly shade the first diagram from 0% to 100%. This represents a regular price.

1b. Erase shading at the top to show a 20% discount.

1c. What percent of the regular price is the shaded (sale) price?

____

2a. Lightly shade the second diagram from 0% up to the 100% mark. This represents another regular price.

2b. How would you use shading to show a 40% markup?

____

____

Use lined paper.

Suppose you have \$250.00 in a bank account earning simple interest, $I = 250 \cdot r \cdot t$, where I is your interest earned, r is the decimal interest rate, and t is the number of years. After 2 years, your money has earned \$7.50 in simple interest.

3. Construct a table showing the simple interest for 1, 2, 3, 4, and 5 years.

4. Find the interest rate that the bank pays.

____

5a. How much simple interest would you earn if you left your \$250.00 in the bank for 20 years?

____

5b. Explain or justify your answer.

____

6. The long-term average gain of the stock market is about 10% per year. How much might your \$250.00 have earned after 20 years invested in the stock market? Treat this like simple interest.

____

Name ______________________ Date ____________ Class ____________

Focus on CAHSEE Standards: Intervention

STANDARD 7NS1.7

Sample Problems

Grade 7 • Number Sense 1.7

Solve problems that involve discounts, markups, commissions, and profit and compute simple and compound interest.

1. The wholesale price of a shirt was $25.00. The retail price is $50. Find the markup.

Markup
= Retail Price − Wholesale Price

Markup = 50.00 − ______ = ______

Markup = ______

2. The wholesale price of a pair of shoes is $35. The retail price is $45. Find the percent markup.

Percent markup = $\frac{\text{Markup}}{\text{Wholesale price}}$

Markup = 45.00 − ______ = ______

Percent markup

= $\frac{\_\_\_\_\_\_}{35.00}$ ≈ ________ ≈ ______

3. The retail price of a graphing calculator is $46.00. The sale price is $34.50. Find the percent discount.

Discount = Retail Price − Sale Price

Discount

= ______ − ______ = ______

Percent discount = $\frac{\_\_\_\_\_\_}{\text{Retail Price}}$

Percent Discount

= $\frac{\_\_\_\_\_\_}{\_\_\_\_\_\_}$ = ________ = ______

4. Alberto makes a $900.00 sale with 20% commission. Find the amount he earns in commission.

Commission
= Sale Amount · Commission Rate

Commission

= ________ · ______ = ________

Commission = __________

5. Leo borrowed $950 at 6% simple interest for 2 years so that he could publish a magazine. If he makes no payments, how much interest will he owe at the end of the 2 years?

What is the problem asking you to find?

____

____

To calculate simple interest, you multiply the principal, the rate of interest per year, and the time in years that the money earns interest. $I = P \cdot r \cdot t$

What is the principal P? __________

What is the rate of interest per year? ____

Over how much time will the interest grow? ________

Substitute these values into the formula. Use a decimal for the rate of interest.

I = ______ · ______ · _____

Multiply. I = ______

The simple interest is __________.

Name ______________________________ Date ____________ Class ____________

STANDARD **7NS1.7**

Focus on CAHSEE Standards: Intervention

Skills Practice

Grade 7 • Number Sense 1.7

Solve problems that involve discounts, markups, commissions, and profit and compute simple and compound interest.

Find the markup.

1. The wholesale price of a pair of gloves is $55.00. The retail price is $75.00.

   The markup is ________.

2. The wholesale price of a soccer ball $44.75. The retail price is $69.99.

   The markup is ________.

Find the percent markup.

3. The wholesale price of a DVD player is $75.00. The retail price is $90.00.

   The percent markup is ______.

4. The wholesale price of a TV is $150.00. The retail price is $225.00.

   The percent markup is ______.

Find the discount.

5. The retail price of a pair of shoes is $55.00. The sale price is $40.00.

   The discount is ________.

6. The retail price of a computer is $1,299.00. The sale price is $1,100.00.

   The discount is __________.

Find the percent discount.

7. The retail price of an office chair is $200.00. The sale price is $150.00.

   The percent discount is ______.

8. The retail price of a digital camera is $300.00. The sale price is $250.00.

   The percent discount is ______.

Find the simple interest.

9. Deposit = $1,500.00
   Interest rate= 5% for 1 year

   Simple interest = ________

10. Loan = $5,750.00
    Interest rate = 11% for 2 years

    Simple interest = ___________

11. Loan = $10,500.00
    Interest Rate = 9% for 2 years

    Simple interest = ___________

Find the amount of tip, tax, or commission.

12. A $25.00 meal with 20% tip

    Amount of tip = ________

13. A $75.00 meal with 15% tip

    Amount of tip = ________

14. A $50.00 item with 6% sales tax

    Amount of sales tax = ________

15. A $3,000.00 item with 6% sales tax

    Amount of sales tax = __________

16. A $900.00 sale with 20% commission

    Amount of commission = __________

Find the compound interest.

17. Loan = $200.00
    Interest rate = 5% for 2 years

    Compound interest = __________

Name ______________________ Date ____________ Class ____________

STANDARD 7NS1.7

Focus on CAHSEE Standards: Intervention

Problem Solving Practice

Grade 7 • Number Sense 1.7

Solve problems that involve discounts, markups, commissions, and profit and compute simple and compound interest.

1. Vera works in a computer store and is paid 15% commission on her sales. In December, her sales total was $5,220. To the nearest dollar, how much did Vera earn in commission in December?

Step 1: What are you asked to find?

____

____

Step 2: What are the facts?

She works in a computer store.	15% commission
Sales total = $5,220	

Cross off the facts you don't need.

Step 3: Break down the problem.

Sales total = ____________

Rate of commission = ______

Step 4: Solve the problem.

Multiply the sales total by the rate of commission. Write the percent as a decimal.

______ • ______ = ______

Commission = __________

Step 5: Check that your answer makes sense. Find the commission for sales totals $5,000 and $5,500. Your answer should fall between the commissions of these amounts.

2. Richard borrowed $1,500 at 11% simple interest for two years so that he could start a computer repair business. If he makes no payments for two years, how much interest will he owe at the end of the second year?

Step 1: What are you asked to find?

____

____

Step 2: What are the facts?
Cross off the facts you don't need.

Borrowed $1,500	Computer business
11% Simple interest	2-year loan

Step 3: Break down the problem.

Principal (loan amount) = ____________

Rate of interest = ______

Number of years = ______

Step 4: Solve the problem.
Substitute the information into this formula:

Simple interest =
principal • rate of interest • number of years

Simple interest = ______ • ______ • ______

Multiply. ______ • ______ • ______ = ______

Simple interest = __________

Step 5: Check your answer. Find the interest for principal amounts of $1,000 and $2,000 for 2 years. Your answer should fall between the interest of these amounts.

Name ______________________ Date __________ Class __________

Focus on CAHSEE Standards: Intervention

STANDARD 7NS2.1

Sample Problems

Grade 7 • Number Sense 2.1

Understand negative whole-number exponents. Multiply and divide expressions involving exponents with a common base.

Find the product. Write your answer in exponential form.

1. $6^7 \times 6^{-4} = ?$

What is the base of the first number?

______

What is the base of the second number?

______

To multiply powers with the same base, keep the base and __________ the exponents.

What addition sentence will you use to find the sum of the exponents?

______ + ______ =

What is the sum? ______

What is $6^7 \times 6^{-4}$? ______

2. Write 2^{-6} using a positive exponent.

What does the negative exponent indicate?

____

____

____

____

Write 2^{-6} using a positive exponent.

______

Find the quotient. Write your answer in exponential form.

3. $8^4 \div 8^2$

What is the base of the first number?

______

What is the base of the second number?

______

To divide powers with the same base, keep the base and __________ the exponents.

What subtraction sentence will you use to find the difference?

______ − ______ =

What is the difference? ______

What is the quotient of $8^4 \div 8^2$? ______

4. Simplify $9^{-1} \times \left(\frac{3}{5}\right)^2$.

Use the order of operations. Perform the exponents first, and then multiply.

Write 9^{-1} as a fraction. $\frac{1}{\_\_}$

Evaluate. $\left(\frac{3}{5}\right)^2 = \frac{3 \times 3}{5 \times \_\_} = \frac{9}{\_\_}$

Simplify. $\frac{1}{\_\_} \times \frac{9}{\_\_} = \frac{\_\_}{\_\_}$

Name ______________________ Date __________ Class __________

Focus on CAHSEE Standards: Intervention

STANDARD 7NS2.1

Skills Practice

Grade 7 • Number Sense 2.1

Understand negative whole-number exponents. Multiply and divide expressions involving exponents with a common base.

Find each product. Write your answer in exponential form.

1. $2^7 \times 2^{-5} =$ ______
2. $3^8 \cdot 3^4 =$ ______
3. $5^3 \cdot 5^{-5} =$ ______
4. $7^3 \times 7^5 =$ ______
5. $10^1 \times 10^{-4} =$ ______
6. $5^{10} \cdot 5^{-7} =$ ______
7. $12^9 \times 12^{-8} =$ ______
8. $4^8 \cdot 4^{-1} =$ ______
9. $11^2 \times 11^7 =$ ______
10. $2^3 \cdot 2^{-13} =$ ______
11. $9^9 \cdot 9^7 =$ ______
12. $13^{11} \times 13^{13} =$ ______

Write each without an exponent.

13. $\frac{1}{2^3} =$ ______
14. $\frac{1}{3^4} =$ ______
15. $\frac{1}{9^3} =$ ______
16. $\frac{1}{4^6} =$ ______
17. $\frac{1}{10^5} =$ ______

Find each quotient. Write your answer in exponential form.

18. $\frac{11^4}{11^2} =$ ______
19. $8^9 \div 8^6 =$ ______
20. $\frac{5^7}{5^2} =$ ______
21. $7^8 \div 7^2 =$ ______
22. $9^{-2} \div 9^2 =$ ______
23. $3^{-2} \div 3^{-4} =$ ______
24. $\frac{15^{19}}{15^{11}} =$ ______

Simplify.

25. $2^{-3} \cdot \left(\frac{4}{5}\right)^2 =$ ______
26. $2^{-2} \times \left(\frac{2}{3}\right)^3 =$ ______

Name ______________________________ Date ____________ Class ____________

STANDARD 7NS2.2

Focus on CAHSEE Standards: Intervention

Alternate Strategies

Grade 7 • Number Sense 2.2

Add and subtract fractions by using factoring to find common denominators.

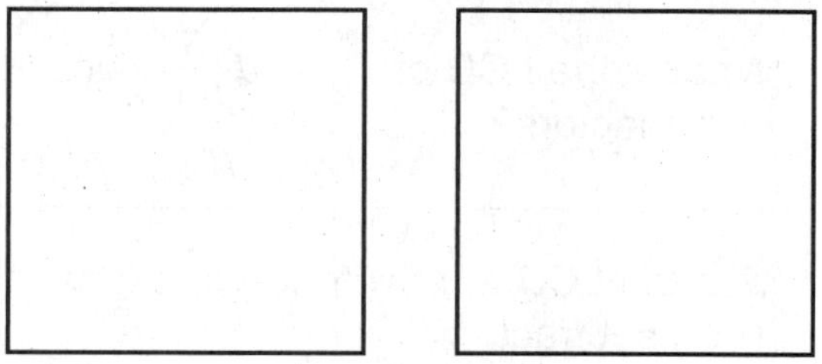

Use the squares.

1a. Draw vertical lines in the first square to make thirds. Shade $\frac{2}{3}$ of the square.

1b. Draw vertical lines in the second square to make fifths. Shade $\frac{3}{5}$ of the square.

2. What addition expression does this model represent? ______________

3a. Draw horizontal lines in both squares to make the same number of parts in each square.

3b. What fractional parts do the small boxes within each square represent?

____

4a. What addition expression does this model represent now?

____

4b. What is the total amount shaded?

____

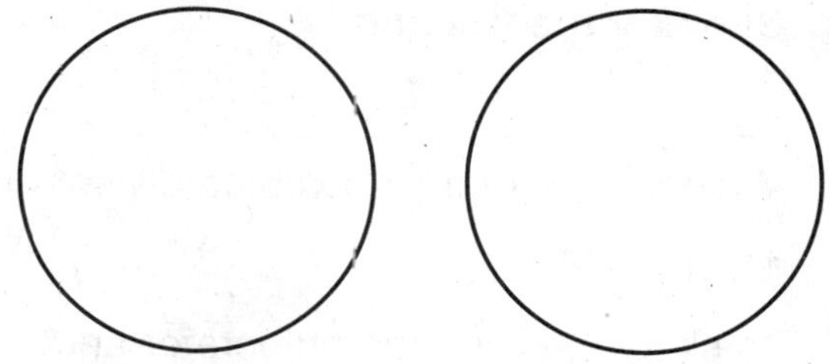

Use the circles.

5a. Shade the first circle. Draw lines in the second circle to make thirds. Shade $\frac{2}{3}$ of that circle.

5b. What mixed number does the model represent? ______________________

6. Suppose you want to use the model to represent subtracting $\frac{3}{4}$. What is the least common multiple of 3 and 4? ____________

7. Draw lines to divide the circles into twelfths. Now what mixed number do the circles represent?

____

8a. What fraction uses 12 as a denominator and is equivalent to $\frac{3}{4}$? ____________

8b. Cross out that many twelfths. What subtraction expression does the model represent?

____

8c. What is the shaded amount that is left?

____

Name ______________________ Date __________ Class __________

STANDARD 7NS2.2

Focus on CAHSEE Standards: Intervention

Sample Problems

Grade 7 • Number Sense 2.2

Add and subtract fractions by using factoring to find common denominators.

1. Simplify the fraction.

$\frac{15}{25}$

What is the greatest common divisor of

15 and 25? ______

To simplify, divide the numerator and denominator by the greatest common divisor.

$$\frac{15 \div \square}{25 \div \square} = \_\_\_\_$$

2. Add.

$\frac{3}{5} + \frac{5}{7}$

Rewrite the fractions using the least common denominator. Since 5 and 7 are prime numbers, their product is the LCD.

Multiply the denominators.

____

Write equivalent fractions.

$$\left(\frac{3}{5} \cdot \frac{7}{7}\right) + \left(\frac{5}{7} \cdot \frac{5}{5}\right) = \_\_\_\_ + \_\_\_\_$$

Add the numerators. Write the sum over the common denominator. Simplify.

____

3. Subtract.

$\frac{11}{12} - \frac{5}{18}$

Write the denominator of $\frac{11}{12}$ using prime factors.

$$\frac{11}{12} = \frac{11}{\_\_\_\_}$$

Write the denominator of $\frac{5}{18}$ using prime factors.

$$\frac{5}{18} = \frac{5}{\_\_\_\_}$$

What is the LCD of $\frac{11}{12}$ and $\frac{5}{18}$ using prime factors?

____

Use the LCD to rewrite the fractions. Then subtract.

$$\frac{11}{12} = 11 \times \frac{\square}{2 \times 2 \times 3 \times 3} = \frac{\square}{\square}$$

$$\frac{5}{18} = 5 \times \frac{\square}{2 \times 2 \times 3 \times 3} = \frac{\square}{\square}$$

Subtract.

____

4. Evaluate the expression $x + y + z$ when $x = \frac{1}{2}, y = \frac{1}{5}, z = \frac{1}{10}$.

Substitute the values of *x*, *y*, and *z* for the variables.

____ + ____ + ____

What is the least common denominator?

____

Write the equivalent fractions.

$$\left(\frac{1}{2} \times \_\_\_\_\right) + \left(\frac{1}{5} \times \_\_\_\_\right) + \_\_\_\_$$

Add.

____ + ____ + ____ = ____

Use the greatest common divisor to simplify the sum.

____ ÷ ____ = ____

Name ______________________ Date __________ Class __________

STANDARD 7NS2.2 Focus on CAHSEE Standards: Intervention

Skills Practice

Grade 7 • Number Sense 2.2

Add and subtract fractions by using factoring to find common denominators.

Add or subtract. Then simplify, if possible.

1. $\frac{3}{4} - \frac{1}{2} =$ ______

2. $\frac{2}{5} + \frac{1}{10} =$ ______

3. $\frac{1}{3} + \frac{1}{6} + \frac{2}{9} =$ ______

4. $\frac{2}{5} - \frac{1}{5} + \frac{1}{2} =$ ______

5. $\frac{2}{3} - \frac{5}{9} =$ ______

6. $\frac{5}{6} - \frac{3}{18} =$ ______

7. $\frac{5}{7} - \frac{2}{3} =$ ______

8. $\frac{6}{9} + \frac{5}{6} =$ ______

9. $\frac{1}{2} + \frac{8}{9} + \frac{1}{6} =$ ______

10. $\frac{8}{5} - \frac{4}{7} =$ ______

11. $\frac{7}{12} + \frac{6}{7} =$ ______

12. $\frac{7}{12} - \frac{5}{24} =$ ______

Evaluate each algebraic expression.

13. $x + y + z$ when $x = \frac{1}{2}, y = \frac{3}{4}, z = \frac{5}{8}$

$x + y + z =$ ______

14. $x + y + z$ when $x = \frac{1}{5}, y = \frac{1}{10}, z = \frac{9}{20}$

$x + y + z =$ ______

15. $x + y - z$ when $x = \frac{1}{10}, y = \frac{1}{20}, z = \frac{1}{20}$

$x + y - z =$ ______

16. $x + y - z$ when $x = \frac{3}{5}, y = \frac{1}{15}, z = \frac{1}{30}$

$x + y - z =$ ______

17. $x - y - z$ when $x = \frac{3}{7}, y = \frac{1}{21}, z = \frac{1}{3}$

$x - y - z =$ ______

18. $x - y - z$ when $x = \frac{1}{3}, y = \frac{2}{9}, z = \frac{1}{18}$

$x - y - z =$ ______

Find the perimeter of the triangle.

19.

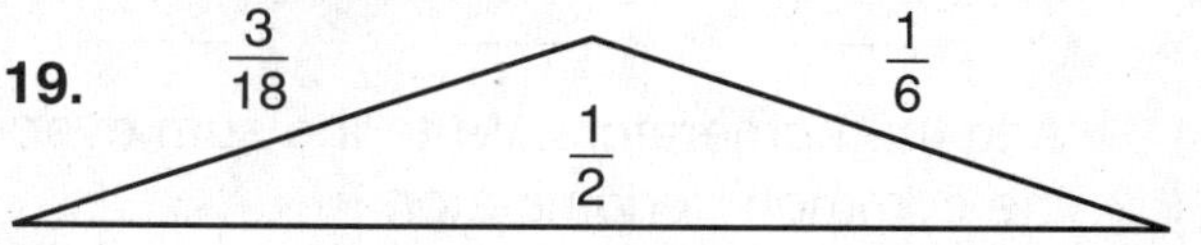

Perimeter = ______ in.

20.

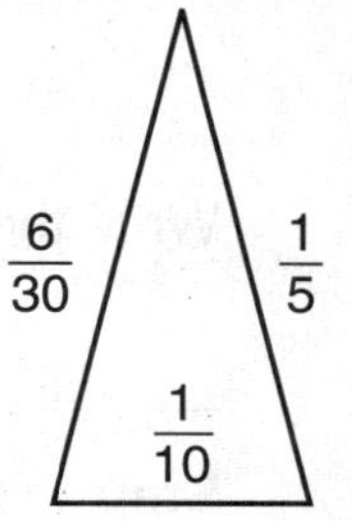

Perimeter = ______ ft

Name ______________________________ Date ____________ Class ____________

STANDARD **7NS2.2**

Focus on CAHSEE Standards: Intervention

Problem Solving Practice

Grade 7 • Number Sense 2.2

Add and subtract fractions by using factoring to find common denominators.

1. Zack used 3 pieces of wire in a project. The wires were $\frac{3}{4}$ in., $\frac{15}{16}$ in., and $\frac{7}{12}$ in. long. How much wire did Zack use?

Step 1: What are you asked to find?

____

Step 2: What are the facts?

~~science project~~ Wires:

____

Cross off the facts you don't need.

Step 3: Break down the problem.

Find the sum: $\frac{3}{4} + \frac{15}{16} + \frac{7}{12}$

Step 4: Solve the problem.

Express denominators as prime factors.

$4 = 2\_\_$ $16 = 2\_\_$ $12 = 2^2 \times \_\_$

Use the prime factors to find the LCD.

$2\_\_ \times \_\_ = \_\_$

Find equivalent fractions with the common denominator.

$\frac{3}{4} = \_\_\_$ $\frac{15}{16} = \_\_\_$ $\frac{7}{12} = \_\_\_$

Add the numerators. Write the sum over the common denominator.

___ + ___ + ___ = ___

Write the answer as a mixed number.

___ = ___ in.

Step 5: Check your answer.

2. Marci bought $\frac{7}{8}$ yd of red fabric and $\frac{5}{6}$ yd of gold fabric. She used $\frac{1}{4}$ yd of the gold fabric. What is the total amount of fabric that she has left?

Step 1: What are you asked to find?

____

Step 2: What are the facts?

____

____

____

Step 3: Break down the problem.

Find the amount left: $\frac{5}{6}$ − ___ + ___

Step 4: Solve the problem.

Use the prime factors to find the LCD.

$2\_\_ \times \_\_ = \_\_$

Find equivalent fractions.

$\frac{7}{8} = \_\_\_$ $\frac{5}{6} = \_\_\_$ $\frac{1}{4} = \_\_\_$

Subtract the amount of gold fabric she used from the amount she bought.

___ − ___ = ___

Add the remaining gold fabric with the amount of red fabric. Simplify.

___ + ___ = ___ = ___

Step 5: Check your answer.

Name ______________________ Date ____________ Class ____________

STANDARD 7NS2.3

Focus on CAHSEE Standards: Intervention

Alternative Strategies

Grade 7 • Number Sense 2.3

Multiply, divide, and simplify rational numbers by using exponent rules.

Simplify $4^2 \cdot 4^3$.

1a. What is 4^2 in expanded form? __________

1b. What is 4^3 in expanded form?

____

1c. Rewrite $4^2 \times 4^3$ in expanded form.

(____________) × (____________)

1d. What is the product of $4^2 \times 4^3$ in simplified form?

____

1e. When you multiply powers with the same base, what should you do with the exponents?

____

Complete each table.

	Product of Powers	Add Exponents	Resulting Power
2.	$2^3 \cdot 2^4$	2^{3+4}	
3.	$3^2 \cdot 3^5$		
4.	$5^6 \cdot 5^2$		

	Product of Powers	Add Exponents	Resulting Power
5.	$y^6 \cdot y^3$		
6.	$a^4 \cdot a^5$		
7.	$b^7 \cdot b^4$		

Simplify $\frac{6^8}{6^5}$.

8a. What is 6^8 in expanded form?

____

8b. What is 6^5 in expanded form?

____

8c. Rewrite $\frac{6^8}{6^5}$ in expanded form.

____

8d. What is the quotient of $\frac{6^8}{6^5}$ in simplified form?

____

8e. When you divide powers with the same base, what should you do with the exponents?

____

Complete each table.

	Quotient of Powers	Subtract Exponents	Resulting Power
9.	$\frac{4^7}{4^3}$		
10.	$\frac{8^5}{8^2}$		
11.	$\frac{7^6}{7^3}$		

	Quotient of Powers	Subtract Exponents	Resulting Power
12.	$\frac{x^8}{x^3}$		
13.	$\frac{r^7}{r^5}$		
14.	$\frac{z^9}{z^3}$		

Name ______________________ Date __________ Class __________

STANDARD 7NS2.3

Focus on CAHSEE Standards: Intervention

Sample Problems

Grade 7 • Number Sense 2.3

Multiply, divide, and simplify rational numbers using exponent rules.

Simplify.

1. $6^3 \cdot 6^4 = ?$

What is the base of the first term? ______

What is the base the second term? ______

To multiply two powers with the same base, add the exponents.

What is the exponent of the first term?

______

What is the exponent of the second term?

______

What is the sum of the two exponents?

______

$6^3 \cdot 6^4 = 6^{3+4} =$ ______

2. $\frac{9^5}{9^2} = ?$

What is the base of the numerator?

______

What is the base of the denominator?

______

To divide two powers with the same base, subtract the exponent of the denominator from the exponent of the numerator.

What is the exponent of the numerator?

______

What is the exponent of the denominator?

______

What is the value of the exponent of the denominator subtracted from the exponent of the numerator? ______

$\frac{9^5}{9^2} = 9^{5-2} =$ ______

3. $-5^2 \cdot (-5)^2 = ?$

What is the base in -5^2? ______

Write -5^2 in expanded form.

______

What is the base in $(-5)^2$? ______

Write $(-5)^2$ in expanded form.

______

Write the product in expanded form.

______

$-5^2 \cdot (-5)^2 =$ ______

4. $\left(\frac{4}{5}\right)^2 = ?$

The exponent outside the parentheses shows that it is part of both the numerator and the denominator.

What is 4^2? ______ What is 5^2? ______

$\left(\frac{4}{5}\right)^2 =$ ______

5. $\frac{12y^6}{18y^4} = ?$

When dividing variable expressions, simplify the whole number and subtract the exponents.

$\frac{12}{18} = \frac{12 \div 6}{18 \div 6} =$ ______

$\frac{y^6}{y^4} = y^{6-4} =$ ______

$\frac{12y^6}{18y^4} =$ ______

Name ______________________ Date ____________ Class ____________

STANDARD 7NS2.3

Focus on CAHSEE Standards: Intervention

Skills Practice

Grade 7 • Number Sense 2.3

Multiply, divide, and simplify rational numbers by using exponent rules.

Multiply the powers with the same base.

1. $6^5 \cdot 6^3$ ________

2. $x^7 \cdot x^3$ ________

3. $-5^6 \cdot -5^3$ ________

4. $8^9 \cdot 8^5$ ________

5. $y^4 \cdot y^7$ ________

6. $-7^2 \cdot -7^4$ ________

7. $-n^2 \cdot -n^3$ ________

Divide the powers with the same base.

8. $\frac{7^5}{7^3}$ ________

9. $\frac{x^9}{x^2}$ ________

10. $\frac{11^6}{11^3}$ ________

11. $\frac{9^9}{9^3}$ ________

12. $\frac{y^8}{y^4}$ ________

13. $\frac{6^7}{6^2}$ ________

14. $\frac{5^5}{5^2}$ ________

Name ____________________ Date __________ Class __________

Focus on CAHSEE Standards: Intervention

STANDARD 7NS2.3 *Skills Practice continued*

Grade 7 • Number Sense 2.3

Multiply, divide, and simplify rational numbers using exponent rules.

Simplify.

15. $\left(\frac{5}{9}\right)^2$ ________

16. $\left(\frac{x}{y}\right)^2$ ________

17. $\frac{5^2}{9}$ ________

18. $\left(\frac{4}{7}\right)^2$ ________

19. $\left(\frac{x}{y}\right)^3$ ________

20. $\frac{3^3}{10}$ ________

21. $\frac{5^2}{12}$ ________

Simplify.

22. $\frac{9xy^5}{3xy^3}$ ________

23. $\frac{25y^7}{50y}$ ________

24. $\frac{15x^6y^7}{3x^3y^5}$ ________

25. $\frac{12x^4y^3}{4xy^3}$ ________

26. $\frac{17x^5}{51x}$ ________

27. $\frac{21x^6y^5}{7x^2y^3}$ ________

28. $\frac{5x^6y^6}{20x^6y^5}$ ________

Name ______________________ Date __________ Class __________

STANDARD 7NS2.4

Focus on CAHSEE Standards: Intervention

Sample Problems

Grade 7 • Number Sense 2.4

Use the inverse relationship between raising a power and extracting the root of a perfect square integer; for an integer that is not a square, determine without a calculator the two integers between which its square root lies and explain why.

1. Which is greater: $\sqrt{450}$ or 21?

Taking the square of a number is the inverse of squaring the number. So, you can square 21 and then compare the product to 450.

What multiplication sentence can you use to find 21^2?

____

What is the product? __________

Which is greater: $\sqrt{450}$ or 21? __________

2. $\sqrt{225} = ?$

The square root of a perfect square is a whole number.

225 is a perfect square.

You can also express $\sqrt{225}$ as the product of two smaller perfect squares.

What are two factors of 225 that are both perfect squares?

____

So, $\sqrt{225} = \sqrt{9} \times \sqrt{25}$

Rewrite the equation, expressing $\sqrt{9}$ and $\sqrt{25}$ as whole numbers.

____

What is $\sqrt{225}$?

____

3. Between which two integers does $\sqrt{17}$ lie?

The square root of a perfect square is a whole number.

$\sqrt{17}$ is not a whole number, 17 is not a perfect square.

Since 17 is not a perfect square, estimate the value of $\sqrt{17}$ by locating it between two perfect squares.

Complete the table.

PERFECT SQUARES		
2^2	=	4
3^2	=	
4^2	=	
5^2	=	

Between which two numbers in the list does 17 lie?

____

Use this information to estimate the value of $\sqrt{17}$.

____

Since 17 is closer to 16 than to 25,

$\sqrt{17}$ is closer to ______ than to 5.

$\sqrt{17} \approx 4$.

Name ______________________ Date __________ Class __________

Focus on CAHSEE Standards: Intervention

STANDARD 7NS2.4

Skills Practice

Grade 7 • Number Sense 2.4

Use the inverse relationship between raising a power and extracting the root of a perfect square integer; for an integer that is not a square, determine without a calculator the two integers between which its square root lies and explain why.

1. $\sqrt{36} =$ ________

2. $\sqrt{169} =$ ________

3. $\sqrt{144} =$ ________

4. $\sqrt{400} =$ ________

5. $\sqrt{196} =$ ________

6. $\sqrt{900} =$ ________

7. $\sqrt{625} =$ ________

8. $\sqrt{289} =$ ________

9. $\sqrt{441} =$ ________

10. $\sqrt{2{,}500} =$ ________

11. **Which is greater: $\sqrt{375}$ or 19?** ________

12. **Which is greater: $\sqrt{121}$ or 12?** ________

13. **Which is greater: $\sqrt{480}$ or 22?** ________

14. **Which is greater: $\sqrt{400}$ or 21?** ________

15. **Which is greater: $\sqrt{1{,}000}$ or 35?** ________

16. **Between which two whole numbers does $\sqrt{24}$ lie?** ________

17. **Between which two whole numbers does $\sqrt{11}$ lie?** ________

18. **Between which two whole numbers does $\sqrt{30}$ lie?** ________

Name ______________________ Date __________ Class __________

Focus on CAHSEE Standards: Intervention

STANDARD 7NS2.5

Alternative Strategies

Grade 7 • Number Sense 2.5

Understand the meaning of absolute value of a number; interpret the absolute value as the distance of the number from zero on a number line; and determine the absolute value of real numbers.

Use a number line to place these numbers in order from least to greatest: |−9|, |−3|, |−7|.

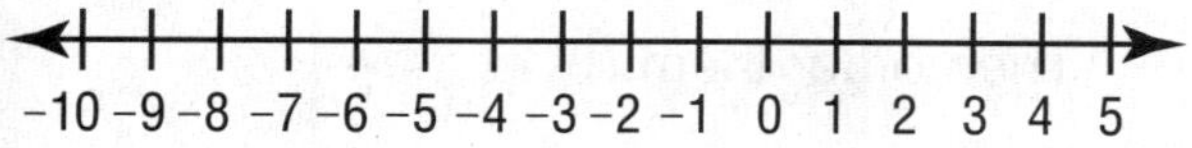

1a. Place a mark on −9, −3, and −7.

1b. How many spaces are between −9 and 0?

____

1c. How many spaces are between −3 and 0?

____

1d. How many spaces are between −7 and 0?

____

1e. List the numbers from least to greatest distance to zero.

____

1f. The absolute value of a number is the distance between that number and zero on a number line. Use the number line to list the absolute values from least to greatest.

____

Use a number line to evaluate |10 − 4|.

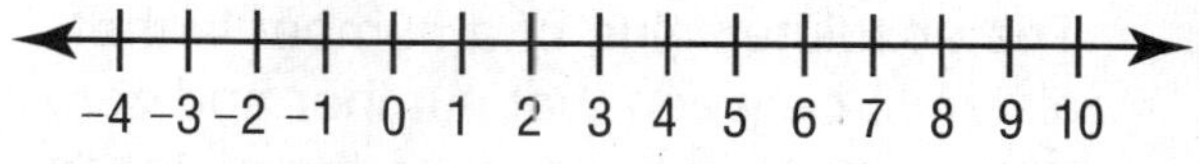

2a. Place a mark on 10 on the number line.

2b. To subtract 4 from 10, move 4 spaces left from 10.

2c. Place a mark on the number line to show the difference of 10 and 4. Then count the number of spaces from 0.

2d. So |10 − 4| = ________.

Evaluate |10 − 4| − |5 − 7| =

3a. Find the value of the numbers inside the first pair of absolute value signs.

____

3b. Find the value of the numbers inside the second pair of absolute value signs.

____

3c. Substitute these values inside the absolute values signs.

____

3d. Find the absolute value of each number and carry out the operation.

____

Name ______________________________ Date ____________ Class ____________

Focus on CAHSEE Standards: Intervention

STANDARD 7NS2.5 *Sample Problems*

Grade 7 • Number Sense 2.5

Understand the meaning of absolute value of a number; interpret the absolute value as the distance of the number from zero on a number line; and determine the absolute value of real numbers.

1. Evaluate |7| and |−7|.

The absolute value of a number is the distance between that number and zero on a number line. Because the distance cannot be negative, the absolute value of a number cannot be negative.

How many spaces are between 7 and 0 on a number line?

____

What is |7|? ________

How many spaces are between −7 and 0 on a number line? ________

What is |−7|? ________

2. Evaluate |−5 + 2|.

If there is an operation sign inside the absolute value signs, evaluate the expression. Then find the absolute value of the number.

What is −5 + 2? ________

What is |−3|? ________

3. Evaluate |−5| + |2|.

If there is an operation sign between two absolute values, find the absolute values first. Then carry out the operation.

What is |−5|? ________

What is |2|? ________

What is 5 + 2? ________

4. Place these numbers in order from least to greatest: |−11|, |−25|, |10|.

Find the absolute value of each number.

Then order them.

What is |−11|? ________

What is |−25|? ________

What is |10|? ________

Order the numbers.

____

5. What is the opposite of 35?

Two numbers that have the same absolute values but different signs are called opposites.

What number has the same absolute value of 35?

____

6. Which expression has the *greater* value, |−25| or |17| ?

|−25| is 25 units from 0, |17| is 17 places from zero. |−25| is more units from 0, and has a greater absolute value.

Name ______________________ Date ____________ Class ____________

STANDARD 7NS2.5

Focus on CAHSEE Standards: Intervention

Skills Practice

Grade 7 • Number Sense 2.5

Understand the meaning of absolute value of a number; interpret the absolute value as the distance of the number from zero on a number line; and determine the absolute value of real numbers.

Find the opposite of each integer.

1. 55

______

2. -70

______

3. -9

______

4. 38

______

5. -47

______

Evaluate.

6. $|-11 + 8|$

______

7. $|21 - 6|$

______

8. $|-3 + 9|$

______

9. $|-7 + -5|$

______

10. $|2 - 9|$

______

Evaluate.

11. $|-20 + 10| - |-7|$

______

12. $|-9 - 5| + |7 - 12|$

______

13. $|-23 + -7| - |-12 - 5|$

______

14. $|-5 + 2| + |2|$

______

15. $|8 + -1| - |-3 + 6|$

______

Place the integers in order from *least* to *greatest*.

16. $|-5|, |0|, |2|$

______

17. $|-27|, |9|, |-11|$

______

18. $|-18|, |3|, |-12|$

______

19. $|17|, |-22|, |6|$

______

20. $|-1|, |-20|, |-11|$

______

Name ______________________ Date __________ Class __________

Focus on CAHSEE Standards: Intervention

STANDARD 7NS2.5

Skills Practice continued

Grade 7 • Number Sense 2.5

Understand the meaning of absolute value of a number; interpret the absolute value as the distance of the number from zero on a number line; and determine the absolute value of real numbers.

Find the opposite of each integer.

1. 36 ________
2. −14 ________
3. 43 ________
4. −20 ________
5. −53 ________

Evaluate.

6. $|-100 + 52|$ ________
7. $|28 - 35|$ ________
8. $|-15 + 21|$ ________
9. $|-8 - 13|$ ________
10. $|25 - 8|$ ________

Evaluate.

11. $|-8 + 15| - |21|$ ________
12. $|-22 - 17| + |3 - 13|$ ________
13. $|-3 - 12| - |-8 + 12|$ ________
14. $|-4 + 9| - |-3 + 8|$ ________
15. $|13 + -6| - |-2 + 11|$ ________

Place the integers in order from *greatest* to *least.*

16. $|-12|, |4|, |-1|$ ________
17. $|-7|, |-21|, |3|$ ________
18. $|-19|, |5|, |12|$ ________
19. $|8|, |-2|, |16|$ ________
20. $|1|, |-94|, |-87|$ ________

Name ______________________________ Date ____________ Class ____________

STANDARD 7AF1.1

Focus on CAHSEE Standards: Intervention

Sample Problems

Grade 7 • Algebra and Functions 1.1

Use variables and appropriate operations to write an expression, an equation, an inequality, or a system of equations or inequalities that represents a verbal description (e.g., three less than a number, half as large as area A).

1. The sum of a number *n* and 15 is 68. Write an equation that shows this relationship.

What variable is the first term you are given? ____

What mathematical operation will result in a sum? ____________

What two terms are being added?

____

What number represents the sum? ____

Complete the equation:

____ + ____ = ____

2. Write an inequality that represents the statement, "A number *x*, decreased by 12, is less than 56."

What variable is the first term you are given? ____

What mathematical operation are you performing when you decrease by a number? ____________

What number is being subtracted from the first term? ____

What symbol shows that one quantity is less than another quantity? ____

Complete the inequality.

____ − ____ < ____

3. A grocer has *y* cases of energy bars in stock. He sells 8 cases and then receives a new shipment containing 3*z* cases. Write an expression that represents the number of cases of energy bars he has in stock after selling some and receiving some cases.

What variable represents the number of cases of energy bars the grocer starts with? ____

When you sell a product, are you adding to your supply or subtraction from your supply? ____________

How many cases does the grocer sell?

____

When you receive a product, are you adding to your supply or subtracting from your supply? ____________

What term represents the number of cases the grocer receives in the new shipment? ____

Complete the expression that represents the number of cases of energy bars the grocer has after selling some and receiving some.

____ − ____ + ____

Name ______________________ Date ____________ Class ____________

STANDARD 7AF1.1

Focus on CAHSEE Standards: Intervention

Skills Practice

Grade 7 • Algebra and Functions 1.1

Use variables and appropriate operations to write an expression, an equation, an inequality, or a system of equations or inequalities that represents a verbal description (e.g., three less than a number, half as large as area A).

Write an equation to show the relationship.

1. The sum of a number *(p)* and 21 is 55.

____

2. The difference between a number *(r)* and 78 is 16.

____

3. The sum of a number *(g)* and 41 is 59.

____

4. The product of a number *(q)* and 7 is 63.

____

5. The product of a number *(d)* and 10 is 50.

____

Write an inequality to represent the statement.

6. A number *m,* multiplied by 15 is greater than 90.

____

7. A number *t,* increased by 17 is less than 43.

____

Write an inequality to represent the statement.

8. A number *b,* multiplied by 13 is less than or equal to 95.

____

9. A number *s,* increased by 21 is less than 39.

____

10. A number *c,* decreased by 41 is greater than 65.

____

11. A number *n,* multiplied by 6 is greater than or equal to 63.

____

Write an expression to represent each situation.

12. In the morning, Sonya collected *h* seashells at the beach. At lunchtime, she gave her brother Sid 7 of her seashells. After lunch, Sonya found another 2*j* shells for her collection. Write the expression that represents the number of seashells Sonya now has.

____

13. A number *v,* increased by 11 is then decreased by 3*w*.

____

Name ______________________________ Date ____________ Class ____________

STANDARD 7AF1.2

Focus on CAHSEE Standards: Intervention

Sample Problems

Grade 7 • Algebra and Functions 1.2

Use the correct order of operations to evaluate algebraic expressions such as $3(2x + 5)^2$.

1. If $x = 2$ and $y = 4$, then $\frac{(xy + 6)}{2} - 3 =$

Substitute for x and y:

$\frac{(\square\,\square + 6)}{2} - 3 =$

Using the order of operations, first evaluate the operations in parentheses:

Multiply:

$\frac{(\square + 6)}{2} - 3 =$

And add:

$\frac{(\square)}{2} - 3 =$

Finally, divide:

____ $- 3 =$

And subtract:

____

2. If $x = 5$ and $y = 4$, then $xy - 2x =$

Substitute for x and y:

____ ____ $- 2$____ $=$

Multiply:

____ $-$ ____ $=$

Then subtract:

____

3. If $x = 3$, then $3(2x - 4)^2 =$

Substitute for x:

$3(2$____ $- 4)^2 =$

Using the order of operations, first evaluate the operations in parentheses:

Multiply inside the parentheses:

$3($____ $- 4)^2 =$

And subtract inside the parentheses:

$3($____$)^2 =$

Next, evaluate the exponent (power).

$3 \bullet$ ____ $=$

Finally, multiply.

____

4. If $x = 4$ and $y = 3$, then $2x(4y - 10)^2 =$

Substitute for x and y:

2____ $(4$____ $- 10)^2 =$

Multiply inside the parentheses, then subtract:

2____ ____$^2 =$

Evaluate the exponent: ____$^2 =$ ____

Multiply: 2____ ____ $=$ ____

Name ______________________ Date ____________ Class ____________

STANDARD **7AF1.2**

Focus on CAHSEE Standards: Intervention

Skills Practice

Grade 7 • Algebra and Functions 1.2

Use the correct order of operations to evaluate algebraic expressions such as $3(2x + 5)^2$.

Show your work for solving each problem.

1. If $x = 5$ and $y = 4$, then $\frac{(xy + 2)}{2} - 1 =$

_____

2. If $a = 3$ and $b = 2$, then $\frac{(ab + 4)}{2} - 2 =$

_____

3. If $y = 3$ and $z = 4$, then $yz - 2z =$

_____

4. If $m = 3$ and $n = 5$, then $mn - 2n =$

_____

5. If $h = 5$, then $2(2h - 7)^2 =$

_____

6. If $v = 2$, then $3(2v - 1)^2 =$

_____

Name ______________________________ Date ___________ Class ___________

STANDARD 7AF1.5

Focus on CAHSEE Standards: Intervention

Sample Problems

Grade 7 • Algebra and Functions 1.5

Represent quantitative relationships graphically and interpret the meaning of a specific part of a graph in the situation represented by the graph.

1. This Venn diagram shows several activities a group of students participated in this winter.

Winter Activities

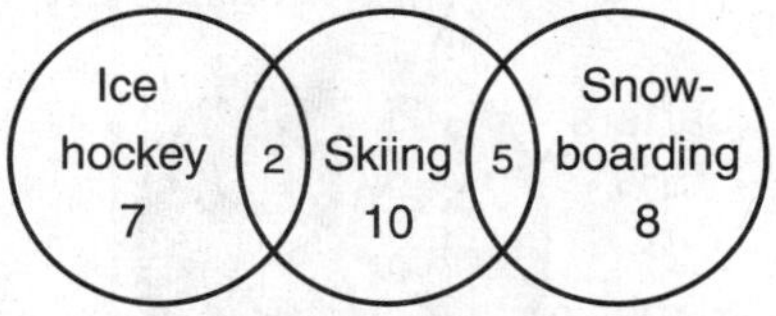

How many students participated in both skiing and snowboarding?

How many students participated in skiing?

2 + ______ + 5 = ______

How many students participated in snowboarding?

______ + 8 = 13

How many students participated in both skiing and snowboarding?

______

2. On an airline flight from Atlanta to Los Angeles, the crew served 160 dinners. The graph shows the dinners served.

Meal Served

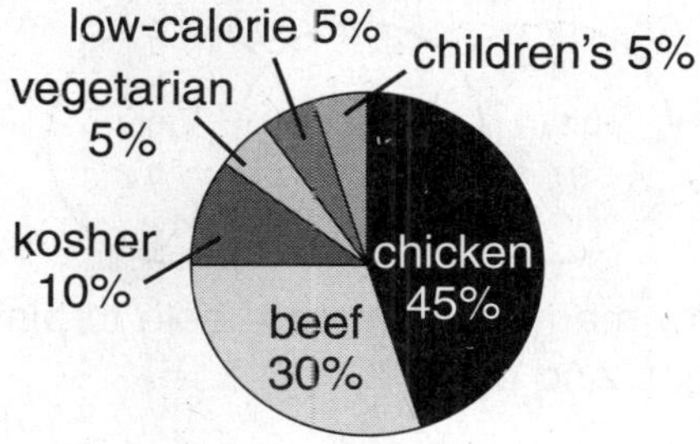

How many more chicken meals than beef meals were served?

How many meals were served in all?

______

What percentage of the meals served were chicken?

______

Find 45% of 160.

160 • ______ = ______

How many chicken meals were served?

______

What percentage of the meals served were beef?

______

Find 30% of 160.

160 • ______ = ______

How many beef meals were served?

______

How many more chicken meals than beef meals were served?

72 − ______ = ______

Name ____________________ Date __________ Class __________

STANDARD **7AF1.5**

Focus on CAHSEE Standards: Intervention

Skills Practice

Grade 7 • Algebra and Functions 1.5

Represent quantitative relationships graphically and interpret the meaning of a specific part of a graph in the situation represented by the graph.

1. This Venn diagram shows several instruments a group of students can play.

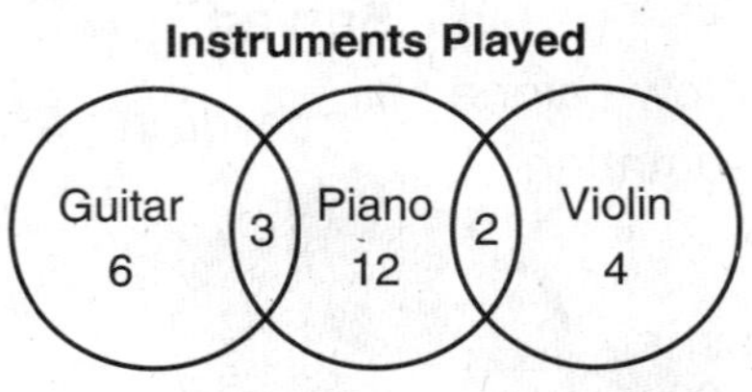

How many students are able to play both guitar and piano?

____

2.

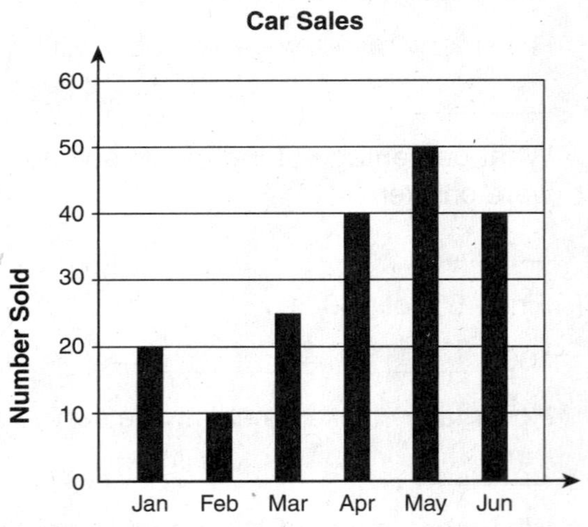

How many more cars were sold in May than in February?

____

3. This circle graph shows the types of coins Erica has in her coin collection.

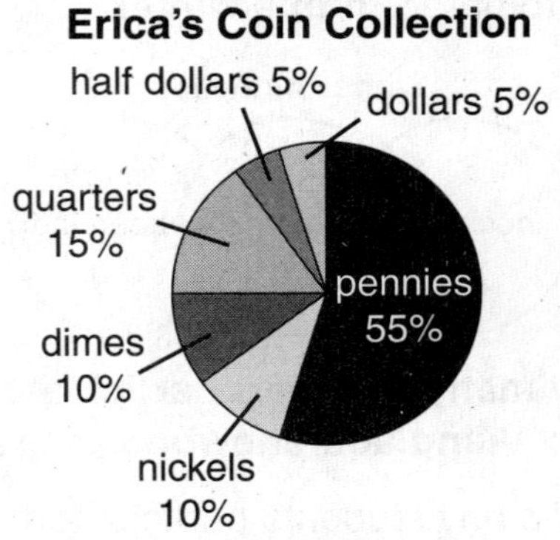

Erica has 240 coins. How many quarters are in the collection?

____

4.

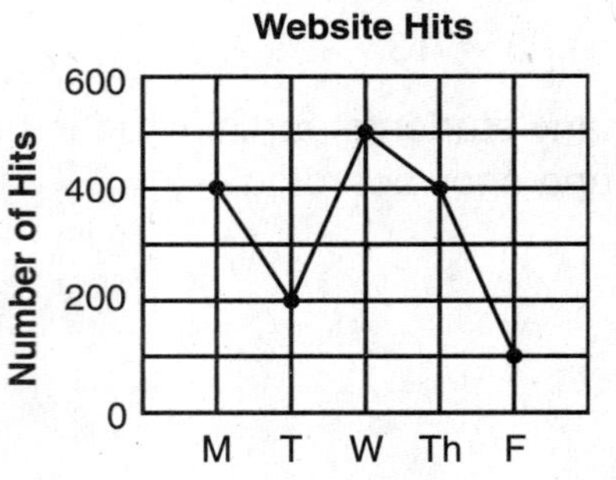

What was the number of Website hits on Wednesday?

____

Name ______________________ Date __________ Class __________

STANDARD **7AF2.1**

Focus on CAHSEE Standards: Intervention

Sample Problems

Grade 7 • Algebra and Functions 2.1

Interpret positive whole-number powers as repeated multiplication and negative whole-number powers as repeated division or multiplication by the multiplicative inverse. Simplify and evaluate expressions that include exponents.

Rewrite in exponential form.

1. $x \cdot x \cdot x$

What operation is being performed?

____

How many factors are there? ____

Are all the factors the same? ____

Rewrite: $x \cdot x \cdot x = x$__

2. xxy

What operation is being performed?

____

How many x factors are there? ____

How many y factors are there? ____

Rewrite: $xxy = x$__ __

3. $2 \cdot x \cdot y \cdot y$

How many 2s are there? ____

How many x factors are there? ____

How many y factors are there? ____

Rewrite: $2 \cdot x \cdot y \cdot y = 2$__ __ __

4. $4xxxyy$

How many 4s are there? ____

How many x factors are there? ____

How many y factors are there? ____

Rewrite: $4xxxyy =$ __x__y__

5. $1 \div x \div x$

$1 \div x = \dfrac{1}{\square}$

$\dfrac{1}{x} \div x = \dfrac{1}{x} \div \dfrac{x}{1}$

$= \dfrac{1}{x} \cdot \dfrac{\square}{\square}$

$= \dfrac{1}{\square^2}$

$= x^{-}$__

6. $2 \div y \div y \div y$

$2 \div y = \dfrac{\square}{y}$

$\dfrac{\square}{y} \div y \div y = \dfrac{2}{y} \cdot \dfrac{1}{\square} \cdot \dfrac{\square}{\square}$

$= \dfrac{2}{\square} =$ ______

Name ______________________ Date ___________ Class ___________

STANDARD
7AF2.1

Focus on CAHSEE Standards: Intervention

Skills Practice

Grade 7 • Algebra and Functions 2.1

Interpret positive whole-number powers as repeated multiplication and negative whole-number powers as repeated division or multiplication by the multiplicative inverse. Simplify and evaluate expressions that include exponents.

Rewrite in exponential form.

1. $y \cdot y \cdot y \cdot y$

____

2. $xyyyyy$

____

3. $3 \cdot x \cdot x \cdot y$

____

4. $4 \cdot x \cdot x \cdot x \cdot y \cdot y$

____

5. $2xxxyyy$

____

6. $8xxxyyyyy$

____

7. $1 \div y \div y$

____

8. $2 \div x \div x$

____

9. $1 \div y \div y \div y$

____

10. $3 \div x \div x \div x \div x$

____

Name ______________________ Date __________ Class __________

Focus on CAHSEE Standards: Intervention

STANDARD 7AF2.2

Sample Problems

Grade 7 • Algebra and Functions 2.2

Multiply and divide monomials; extend the process of taking powers and extracting roots to monomials when the latter results in a monomial with an integer exponent.

1. Simplify $(3a^2b)(4a^3b^2)$.

Expand each term:

$(3a^2b) = 3 \cdot \_\_ \cdot \_\_ \cdot b$

$(4a^3b^2) = 4 \cdot a \cdot \_\_ \cdot \_\_ \cdot b \cdot \_\_$

Multiply the numerals:

$3 \cdot 4 = \_\_\_\_$

Multiply the variables separately:

$a \cdot a \cdot a \cdot a \cdot a = a\_\_$

$b \cdot b \cdot b = b\_\_$

Combine the terms:

$12 \cdot a^5 \cdot b^3 = \_\_\_\_\_\_$

2. Divide $\frac{8x^3}{4x}$.

Expand the numerator and denominator:

$\frac{8x^3}{4x} = \frac{8 \cdot x \cdot \square \cdot \square}{4 \cdot \square}$

Reduce to lowest terms:

$= \frac{\overset{\square\,1}{\cancel{8}} \cdot \cancel{x} \cdot x \cdot x}{\underset{\square\,1}{\cancel{4}} \cdot \cancel{x}}$

Combine the remaining terms:

$2 \cdot x \cdot x = 2x\_\_$

3. Simplify $(2a^4)^2$.

Apply the exponent outside the parentheses to each factor inside:

$(2a^4)^2 = \_\_\_\_^2 \cdot \_\_\_\_^2$

Evaluate each factor:

$2^2 = \_\_\_\_$

$(a^4)^2 = a^{(4 \cdot \_\_)} = a\_\_$

Combine the terms: $4 \cdot a^8 = \_\_\_\_\_\_$

4. Simplify $\frac{(3x^3)^3}{9x^5}$.

Apply the exponent outside the parentheses to each factor inside:

$(3x^3)^3 = \_\_\_\_^3 \cdot \_\_\_\_^3$

Evaluate each factor:

$3^3 = \_\_\_\_$

$(x^3)^3 = x^{(3 \cdot \_\_)} = x\_\_$

Combine terms: $27 \cdot x^9 = \_\_\_\_\_\_$

Expand the numerator and denominator and reduce to lowest terms:

$\frac{27x^9}{9x^5} = \frac{\cancel{27} \cdot \cancel{x} \cdot \cancel{x} \cdot \cancel{x} \cdot \cancel{x} \cdot \cancel{x} \cdot x \cdot x \cdot x \cdot x}{\cancel{9} \cdot \cancel{x} \cdot \cancel{x} \cdot \cancel{x} \cdot \cancel{x} \cdot \cancel{x}}$

Combine the remaining terms:

$3 \cdot x \cdot x \cdot x \cdot x = \_\_\_\_\_\_$

Name ________________________ Date __________ Class __________

STANDARD 7AF2.2

Focus on CAHSEE Standards: Intervention

Skills Practice

Grade 7 • Algebra and Functions 2.2

Multiply and divide monomials; extend the process of taking powers and extracting roots to monomials when the latter results in a monomial with an integer exponent.

1. Simplify $(2xy^3)(6x^2y^2)$.

____

2. Simplify $(7a^5b^4)(4ab)$.

____

3. Simplify $(9x^3y^3z)(3x^3yz)$.

____

4. Divide $\frac{18a^5}{3a^2}$.

____

5. Divide $\frac{10x^4}{5x^3}$.

____

6. Divide $\frac{18a^6}{6a^2}$.

____

7. Simplify $(2x^2y)^3$.

____

8. Simplify $\frac{(3ab^3)^2}{12b^3}$.

____

Name ______________________ Date __________ Class __________

STANDARD 7AF3.1

Focus on CAHSEE Standards: Intervention

Sample Problems

Grade 7 • Algebra and Functions 3.1

Graph functions of the form $y = nx^2$ and $y = nx^3$ and use in solving problems.

1. Use the rule $y = x^2$ to complete the function table. Then graph the function for all integers x, from -3 to 3.

x	y
−3	
−2	
−1	
0	
1	
2	
3	

How can you find the value of y?

____

Will the product of a negative number squared be positive or negative?

____

Plot the points from your function table. Then join the points.

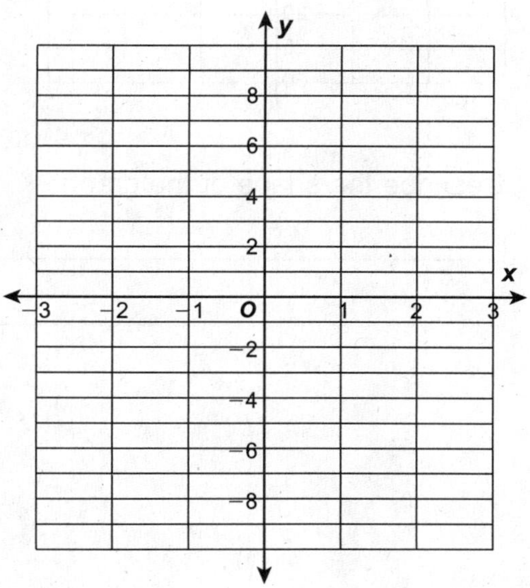

2. Use the rule $y = x^3$ to complete the function table. Then graph the function for all integers x, from -3 to 3.

x	y
−3	
−2	
−1	
0	
1	
2	
3	

How can you find the value of y?

____

____

Will the product of a negative number cubed be positive or negative?

____

Plot the points from your function table. Then join the points.

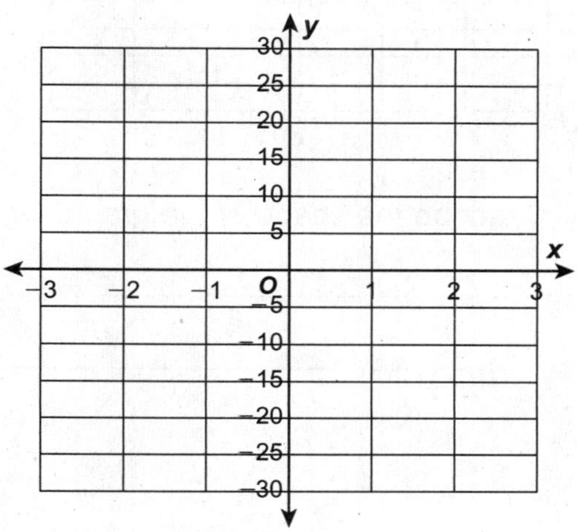

Name ______________________ Date ____________ Class ____________

STANDARD **7AF3.1**

Focus on CAHSEE Standards: Intervention

Skills Practice

Grade 7 • Algebra and Functions 3.1

Graph functions of the form $y = nx^2$ and $y = nx^3$ and use in solving problems.

Complete each function table. Then graph the function.

1. Rule: $y = -x^2$

x	y
−3	
−2	
−1	
0	
1	
2	
3	

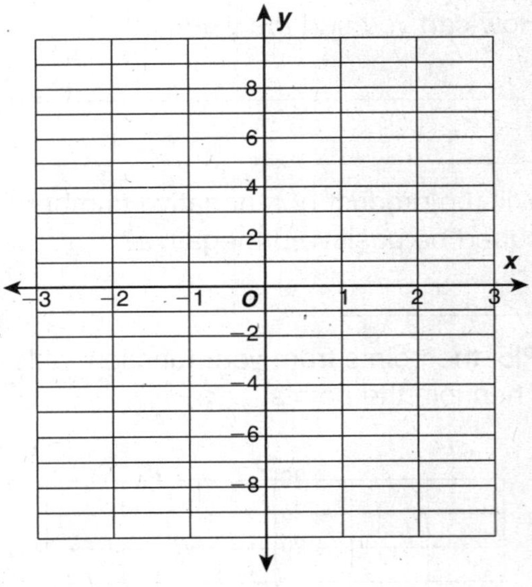

Describe the shape of the graph.

____

____

2. Rule: $y = -x^3$

x	y
−3	
−2	
−1	
0	
1	
2	
3	

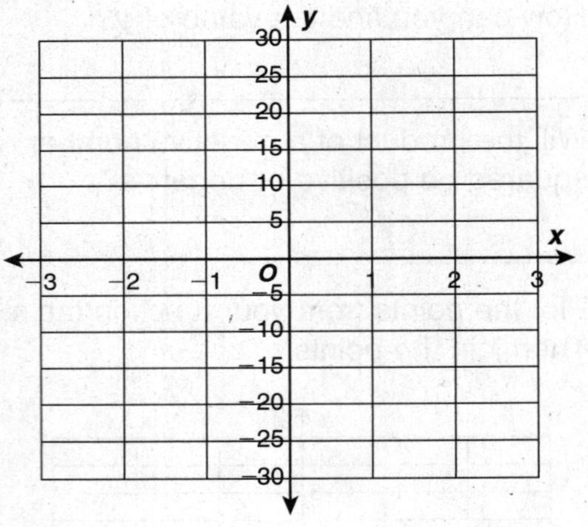

Describe the shape of the graph.

____

____

Name ______________________________ Date ____________ Class ____________

Focus on CAHSEE Standards: Intervention

STANDARD 7AF3.3

Alternative Strategies

Grade 7 • Algebra and Functions 3.3

Graph linear functions, noting that the vertical change (change in *y*-value) per unit of horizontal change (change in *x*-value) is always the same and know that the ratio ("rise over run") is called the slope of a graph.

What is the slope of this line?

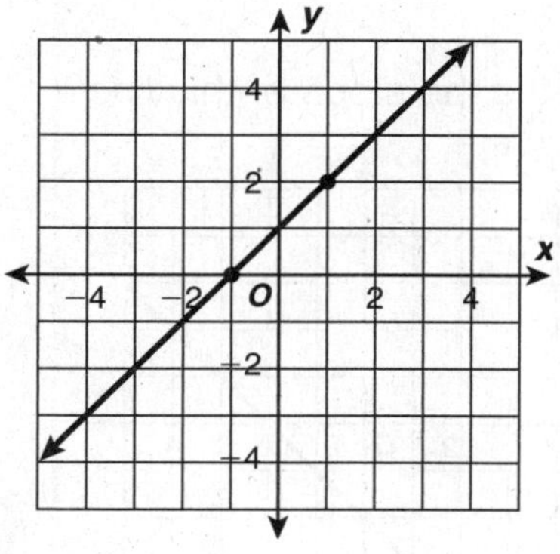

Use the graph.

1a. Using the graph above, start with the lower point and count up to the second point on the *y*-axis (rise). Write this number. _____

1b. Count the number of units to the right (run). How many units to the right did you count? _____

1c. Write this fraction. $\frac{\text{rise}}{\text{run}} =$ _____

1d. Write the fraction in simplest form. _____

1e. What is the slope? _____.

2a. Make your own graph using (2, 1) and (1, –1). What is the rise and run?

____

2b. What is the slope? _____

What is the slope of this line?

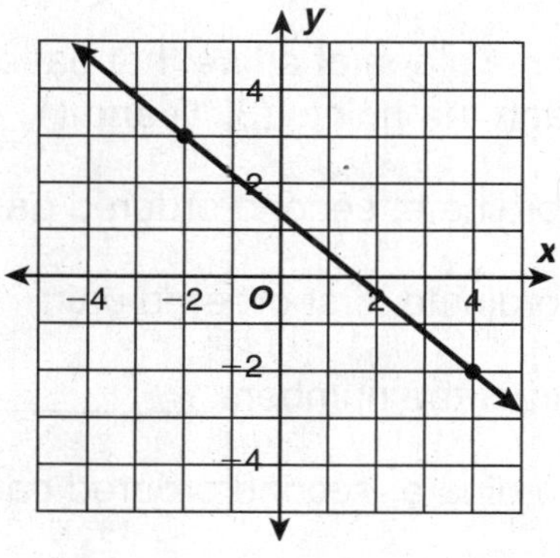

Use the graph.

3a. Using the graph above, start with the lower point and count up to the second point on the *y*-axis (rise). Write this number. _____

3b. Count the number of units to the left of the upper coordinate (run). (Remember that counting to the left will result in a negative number.) How many units to the left did you count? _____

3c. Write this fraction. $\frac{\text{rise}}{\text{run}} =$ _____

3d. What is the slope? _____

4a. Make your own graph using (2, −3) and (−1, 1). What is the rise and run?

____

4b. What is the slope? _____

Name ______________________________ Date ____________ Class ____________

Focus on CAHSEE Standards: Intervention

STANDARD 7AF3.3 *Sample Problems*

Grade 7 • Algebra and Functions 3.3

Graph linear functions, noting that the vertical change (change in *y*-value) per unit of horizontal change (change in *x*-value) is always the same and know that the ratio ("rise over run") is called the slope of a graph.

Solve for slope. Use the formula $\frac{(y_2 - y_1)}{(x_2 - x_1)}$.

1. Find the slope of a line that passes through the points (3, 1) and (1, 4).

   y coordinate, second ordered pair: _____

   y coordinate, first ordered pair: _____

   Subtract the numbers. ______________

   x coordinate, second ordered pair: _____

   x coordinate, first ordered pair: _____

   Subtract the numbers. ______________

   $\frac{\text{change of } y}{\text{change of } x} =$ ________

2. Find the slope of a line that passes through the points (3, 5) and (4, 7).

   y coordinate, second ordered pair: _____

   y coordinate, first ordered pair: _____

   Subtract the numbers. ______________

   x coordinate, second ordered pair: _____

   x coordinate, first ordered pair: _____

   Subtract the numbers. ______________

   $\frac{\text{change of } y}{\text{change of } x} =$ ________

Solve for slope in the coordinate plane below.

3. What is the slope of this line?

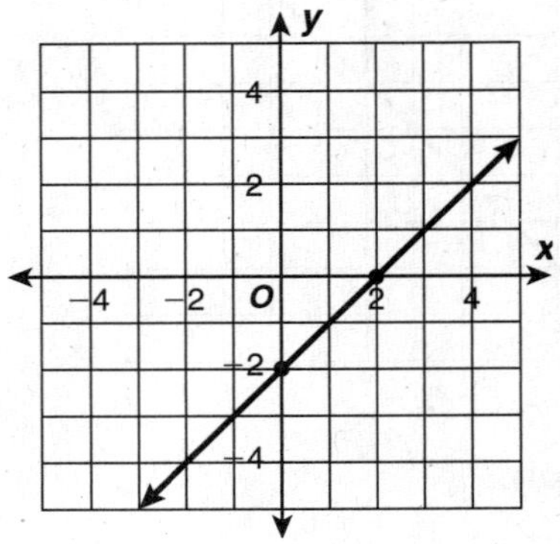

Write the formula to find the slope of a line: slope $= \frac{\text{rise}}{\text{run}}$

The lower coordinate pair is (0, −2). Move up the *y*-axis to reach the upper coordinate pair.

Write the number of units moving up the *y*-axis, or the rise. _____

Next, from where the counting stopped along the *y*-axis, count the units to the right until you reach the upper coordinate pair (2, 0).

Write the number of units or the run. ____

Write the slope of the line.

$\frac{\text{rise}}{\text{run}} =$ _____ $= 1$

Name ______________________________ Date ____________ Class ____________

Focus on CAHSEE Standards: Intervention

STANDARD 7AF3.3 *Skills Practice*

Grade 7 • Algebra and Functions 3.3

Graph linear functions, noting that the vertical change (change in *y*-value) per unit of horizontal change (change in *x*-value) is always the same and know that the ratio ("rise over run") is called the slope of a graph.

Find the slope of each line.

1.

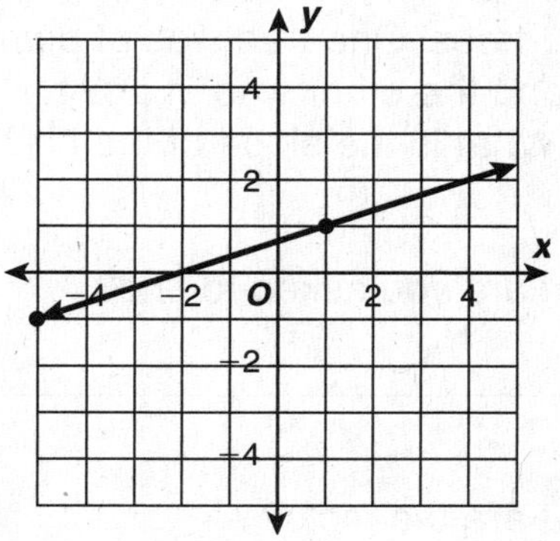

____

2.

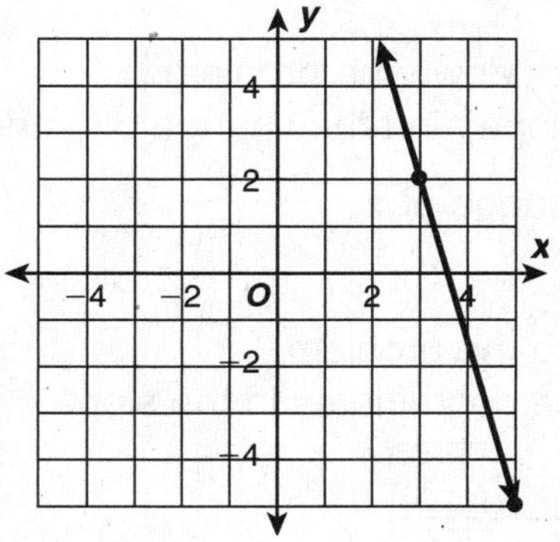

____

3. Find the slope of the line that passes through points (−2, 4) and (4, 2).

____

4. Find the slope of the line that passes through points (3, 5) and (4, 7).

____

5. Find the slope of the line that passes through points (6, 2) and (8, 0).

____

6. Find the slope of the line that passes through points (−2, −2) and (1, 4).

____

7. Find the slope of the line that passes through points (−5, 3) and (7, −6).

____

8. Find the slope of the line that passes through points (−3, 2) and (1, 4).

____

Name ______________________ Date __________ Class __________

STANDARD 7AF3.3

Focus on CAHSEE Standards: Intervention

Problem Solving Practice

Grade 7 • Algebra and Functions 3.3

Graph linear functions, noting that the vertical change (change in *y*-value) per unit of horizontal change (change in *x*-value) is always the same and know that the ratio ("rise over run") is called the slope of a graph.

1. Toni drew a graph of her hike on a coordinate plane. She passed 2 points on her way up, one was (−2, −2), and the other was (3, 2). What is the slope of the hill?

Step 1: What are you asked to find?

____

Step 2: What are the facts?

What is the first ordered pair? __________

What is the second ordered pair? __________

Step 3: Break down the problem.
To find the slope, find the change of *y* and put it over change of *x*. $\frac{(y_2 - y_1)}{(x_2 - x_1)}$

Step 4: Solve the problem.
Substitute the coordinates in the slope formula. $\frac{(2 - (\square))}{(3 - (\square))}$

What is the slope? ______

Step 5: Check that your answer makes sense.

2. Bill drew a graph of his ski run on a coordinate plane. On the way down he passed 2 trees. One tree was at point (4, −1) and the other was at point (−3, 2). What is the slope of the hill?

Step 1: What are you asked to find?

____

Step 2: What are the facts?

What is the first ordered pair? __________

What is the second ordered pair? __________

Step 3: Break down the problem.
To find the slope, find the change of *y* and put it over change of *x*. $\frac{(y_2 - y_1)}{(x_2 - x_1)}$

Step 4: Solve the problem.
Substitute the coordinates in the slope formula. $\frac{(2 - (\square))}{(-3 - \square)}$

What is the slope? ______

Step 5: Check that your answer makes sense.

Name ______________________ Date __________ Class __________

Focus on CAHSEE Standards: Intervention

STANDARD **7AF3.4**

Alternative Strategies

Grade 7 • Algebra and Functions 3.4

Plot the values of quantities whose ratios are always the same (e.g., cost to the number of an item, feet to inches, circumference to diameter of a circle). Fit a line to the plot and understand that the slope of the line equals the quantities.

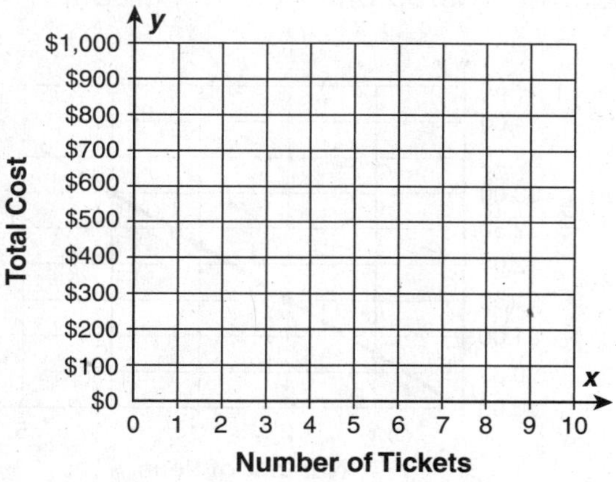

Use the coordinate grid and the unit price ratio.

1a. You are given that 2 tickets cost $150. Plot the point.

1b. What is the unit price per ticket? Use a ratio.

$$\frac{2}{150} = \frac{1}{x}$$

$$2x = 150$$

$x =$ ______

The unit price is ______.

1c. Plot the point. ________

1d. Complete a function table using the unit price per ticket.

Tickets	0	2	4	6	8	10
Cost	0	150				

1e. Plot the points from your function table. Then join the points.

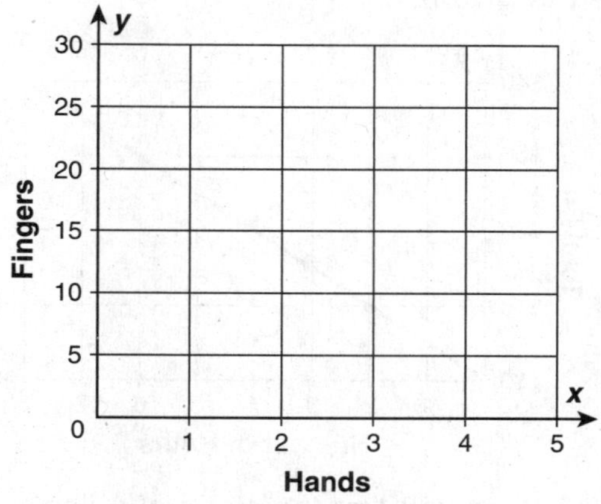

Use the coordinate grid and a ratio.

2a. What is the relationship between the number of hands and the number of fingers?

____

2b. Use the hand-finger ratio to find the number of fingers on 2 hands.

$$\frac{1}{5} = \frac{2}{x}$$

$$x = 10$$

2c. Graph two points using the hand-finger ratio. Find the change in *y*. ______

2d. Using the same two points, find the change in *x*. ______

2e. What is the slope of the line?

____

Name ____________________ Date __________ Class __________

Focus on CAHSEE Standards: Intervention

STANDARD **7AF3.4**

Sample Problems

Grade 7 • Algebra and Functions 3.4

Plot the values of quantities whose ratios are always the same (e.g., cost to the number of an item, feet to inches, circumference to diameter of a circle). Fit a line to the plot and understand that the slope of the line equals the quantities.

Use the graph.

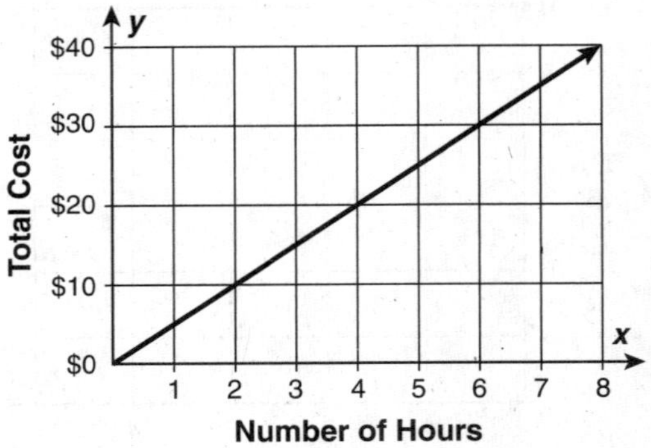

1. How do you find the slope of a line?

____

To find the change in *y*, what do you need to do?

____

____

____

To find the change in *x*, what do you need to do?

____

____

____

Find the change in *y*.

____

Find the change in *x*.

____

What is the slope? _____

What is the ratio of cost to hours? _____

Plot the ordered pairs on this graph.

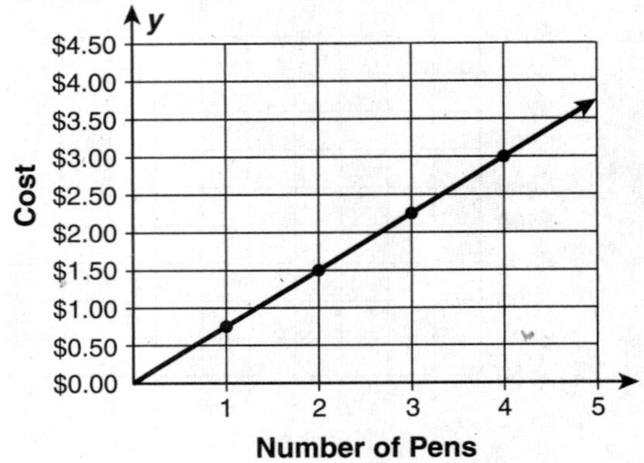

2. Nikki bought 4 pens for $3.00.

What ordered pair can you plot? _______

What is the price for 1 pen? _______

What equation can you write to find the cost of any number of pens?

____

Complete the table.

Pens	0	1	2	3	4	5
Cost $	0					

Plot the points on the graph for each number of pens from 0 to 5.

Name ______________________ Date __________ Class __________

STANDARD 7AF3.4

Focus on CAHSEE Standards: Intervention

Skills Practice

Grade 7 • Algebra and Functions 3.4

Plot the values of quantities whose ratios are always the same (e.g., cost to the number of an item, feet to inches, circumference to diameter of a circle). Fit a line to the plot and understand that the slope of the line equals the quantities.

1. All CDs at *Just CDs* cost the same amount of money. One customer bought 2 CDs for $16 and another customer bought 7 CDs for $56. Use the coordinate grid to show the number of CDs bought from 0 CDs to 10 CDs.

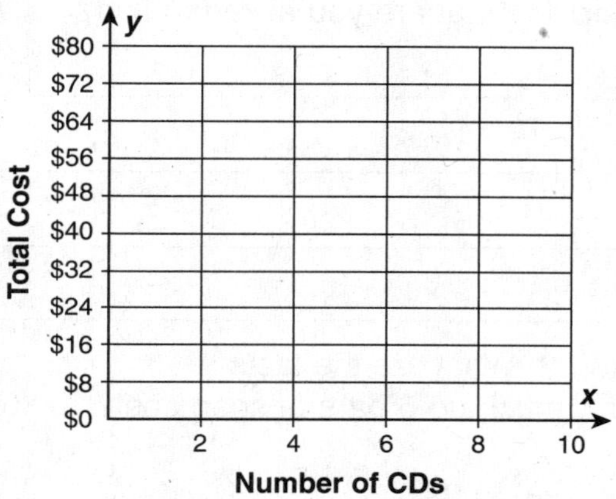

2. Mr. Jenkins bought 4 gallons of gasoline for $9.00. Complete the coordinate grid to show the price for gasoline from 0 gallons to 8 gallons.

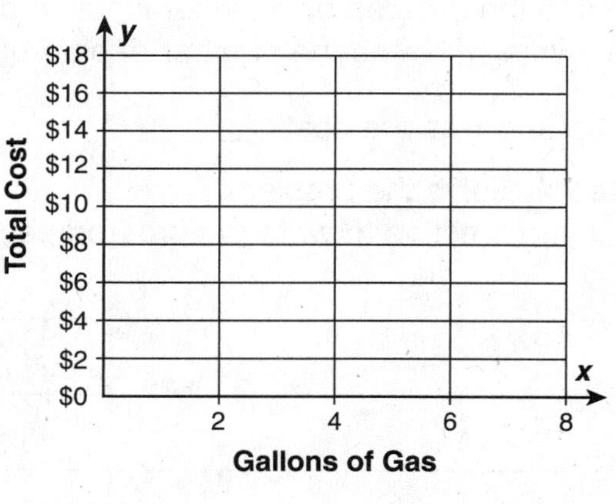

3. Ms. Lee bought 5 apples for $1.50. Complete the coordinate grid to show the relationship between the cost and the number of apples purchased from 0 apples to 10 apples.

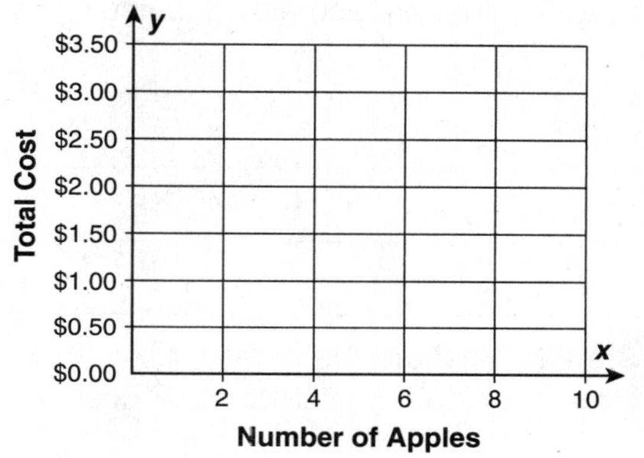

4. Three friends bought tickets for the movies for a total of $21. Complete the coordinate grid to show the relationship between the cost and the number of tickets purchased from 0 tickets to 10 tickets.

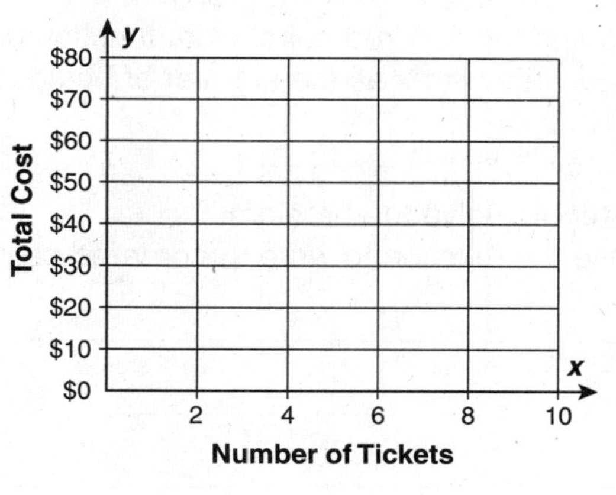

Name ______________________ Date __________ Class __________

STANDARD **7AF3.4**

Focus on CAHSEE Standards: Intervention

Problem Solving Practice

Grade 7 • Algebra and Functions 3.4

Plot the values of quantities whose ratios are always the same (e.g., cost to the number of an item, feet to inches, circumference to diameter of a circle). Fit a line to the plot and understand that the slope of the line equals the quantities.

1. Luigi bought 4 bottles of water for $2.40. Mark bought 9 bottles of water for $5.40. Write the ordered pairs that could be used on a coordinate grid to show the cost of 0–10 bottles of water.

Step 1: What are you asked to find?

____

____

____

____

Step 2: What are the facts?
How much do 4 bottles of water cost?

______

How much do 9 bottles of water cost?

______

What is the unit price per bottle? ______

Step 3: Break down the problem.
To find the ordered pairs, what function can you write with *x* as the number of bottles and *y* as the cost? __________

Step 4: Solve the problem.
Use the function to write the ordered pairs.

____

____

____

Step 5: Check that your answer makes sense.

2. At Smitty's, it does not matter how many bars of soap you buy. All items have the same unit price. Connie bought 6 bars of soap for $7.20. Write the ordered pairs that could be used on a coordinate grid to show the cost of 0 – 10 bars of soap.

Step 1: What are you asked to find?

____

____

____

____

Step 2: What are the facts?
How much do 6 bars of soap cost?

______

What is the unit price per bar of soap?

______

Step 3: Break down the problem.
To find the ordered pairs, what function can you write with *x* as the number of bars of soap and *y* as the cost? __________

Step 4: Solve the problem.
Use the function to write the ordered pairs.

____

____

____

Step 5: Check that your answer makes sense.

Name ______________________________ Date __________ Class __________

Focus on CAHSEE Standards: Intervention

STANDARD 7AF4.1

Alternative Strategies

Grade 7 • Algebra and Functions 4.1

Solve two-step linear equations and inequalities with one variable over the rational numbers, interpret the solution or solutions in the context from which they arose, and verify the reasonableness of the results.

Use integer chips to solve the equation.

1. What is the value of x in the equation below?

$3x + 6 = 12$

1a. Place 6 chips on the left side of the equation and 12 chips on the right side of the equation.

$3x +$ ______ $=$ ______

1b. Subtract 6 chips from each side of the equation to isolate the variable.

$3x +$ ______ $=$ ______

1c. Divide the remaining chips into 3 equal groups. The number 3 is the coefficient.

$3x =$ ______

1d. Write the solution to the equation.

______

1e. Check your answer. $(3 \times 2) + 6 = 12$. The solution is correct.

2. Try it yourself. Create your own equation and solve for the variable.

Use integer chips to solve the inequality.

Let ○ = 1 and ● = −1.

Remember (●○) = 0 or they cancel each other out.

3. What is the value of x in the inequality below?

$2x - 4 \geq 10$

3a. Place 4 gray chips on the left side of the inequality and 10 white chips on the right side of the inequality.

$2x -$ ●● $\geq$ ○○○○○
●● ○○○○○

3b. Add 4 chips to each side of the inequality to isolate the variable.

$2x \geq$ ○○○○○○○
○○○○○○○

3c. Divide the remaining chips into 2 groups. The number 2 is the coefficient.

$2x \geq$ (○○○○○○○)
(○○○○○○○)

3d. Write the solution to the inequality.

______

3e. Check your answer. $(2 \times 7) - 4 \geq 10$. The solution is correct. Graph the solution on a number line.

Name ______________________ Date ____________ Class ____________

STANDARD **7AF4.1**

Focus on CAHSEE Standards: Intervention

Sample Problems

Grade 7 • Algebra and Functions 4.1

Solve two-step linear equations and inequalities with one variable over the rational numbers, interpret the solution or solutions in the context from which they arose, and verify the reasonableness of the results.

Solve the equation.

1. $8x + 25 = 89$

Subtract 25 from both sides of the equation. Then divide both sides by 8 to isolate the variable.

$8x + 25 = 89$

$-\quad -$

$8x =$

$\frac{8x}{} = \_\_$

$x =$

Check by substituting 8 for x.

$8\_\_\_\_ + 25 = 89$

$64 + 25 = 89$ ✓

2. $21x - 15 = 12$

Add 15 to both sides of the equation, which is using the subtraction inverse of equality. $21x - 15 = 12$

$+\quad +$

$21x =$

$\frac{21x}{} = \_\_$

$x =$ ____________

Check:

$21 \times \_\_\_\_ - 15 = 12$

$\_\_\_\_ - 15 = 12$

$27 - 15 = 12$ ✓

Solve the inequality.

3. $2x - 15 > -9$

Add 15 to both sides of the inequality, which is using the subtraction inverse of equality. Then divide both sides by 2 to isolate the variable.

$2x - 15 > -9$

$+\quad +$

$2x >$

$\frac{2x}{} = \_$

$x >$

Graph the solution on a number line.

4. $16 - 10y > 34$

Subtract 16 from both sides of the inequality sign, which is using the additive inverse of equality. Then divide both sides by -10 to isolate the variable.

$16 - 10y > 34$

$-\quad -$

$-10y >$

$\frac{-10y}{} > \_\_$

$y < \_\_\_\_$ or $y < \_\_\_\_\_\_\_\_$

Name ______________________ Date __________ Class __________

STANDARD 7AF4.1

Focus on CAHSEE Standards: Intervention

Skills Practice

Grade 7 • Algebra and Functions 4.1

Solve two-step linear equations and inequalities with one variable over the rational numbers, interpret the solution or solutions in the context from which they arose, and verify the reasonableness of the results.

1. Solve for x.

$$2x + 4 = 14$$

____

2. Solve for x.

$$3x - 9 = 27$$

____

3. Solve for y.

$$3y - 14 = 4$$

____

4. Solve for y.

$$-4y + 6 = -10$$

____

5. Solve for x.

$$2.5x + 5 = 10.5$$

____

6. Find the solution set for y.

$$5y + 12 \geq 17$$

____

7. Find the solution set for x.

$$16 \geq -2x + 2$$

____

8. Find the solution set for x.

$$17 > 21 - 15x$$

____

9. Find the solution set for y.

$$-5 > 15y + 5$$

____

10. Find the solution set for y.

$$3y + 5 < 26$$

____

Name ______________________ Date __________ Class __________

Focus on CAHSEE Standards: Intervention

STANDARD **7AF4.1** *Problem Solving Practice*

Grade 7 • Algebra and Functions 4.1

Solve two-step linear equations and inequalities with one variable over the rational numbers, interpret the solution or solutions in the context from which they arose, and verify the reasonableness of the results.

Solve the equation.

1. Sally wants to fill a decorative bag with fruit for her best friend. She wants to spend $40. The bag costs $5.00 and each piece of fruit costs $0.25. How many pieces of fruit will be in the bag?

Step 1: What are you asked to find?

____

Step 2: What are the facts?
What is the amount of money that Sally will spend? ______

How much does the bag cost? ______
How much does each piece of fruit cost?

______

Step 3: Break down the problem.
Cost of fruit + Cost of bag = Total cost

______x + ____ = ____

Step 4: Solve the equation.
Subtract 5 from both sides of the equation.
Divide both sides of the equation by 0.25.

____

What is x? ______

Step 5: Check your answer.

0.25______ + ____ = ____

____ + 5 = 40

40 = 40 ✓

Solve the inequality.

2. You are going to the movies. It costs you $7.50 to get into the movie theater and $2.50 for every snack that you want. You don't want to spend more than $25. How many snacks can you buy?

Step 1: What are you asked to find?

____

____

Step 2: What are the facts?
How much does it cost to get into the movie theater? ______

How much is each snack? ______
How much money can you spend?

____

Step 3: Break down the problem.
Cost of snacks + Cost of movies ≤ Total cost

______s + ____ ≤ ____

Step 4: Solve the problem.
Subtract 7.5 from both sides of the inequality.
Divide both sides of the inequality by 2.5.

____

What is the solution set for s? ______

Step 5: Check your answer.

2.5____ + ____ ≤ 25

______ + 7.5 ≤ 25

25 ≤ 25 ✓

Name ______________________________ Date __________ Class __________

Focus on CAHSEE Standards: Intervention

STANDARD 7AF4.2

Alternative Strategies

Grade 7 • Algebra and Functions 4.2

Solve multistep problems involving rate, average speed, distance, and time or direct variation.

1. Mr. Ramirez earns $65 for 5 hours of work. At that rate, how many hours would he have to work to earn $455?

Hours	1	2	3	4	5	?
Earnings	?	?	?	?	$65	$455

Find how much Mr. Ramirez earns for 1 hour.

$\frac{\$65}{5} =$ ______

Complete the table.

Hours	1	2	3	4	5	?
Earnings					$65	$455

Find how many hours Mr. Ramirez works to earn $455.

$\frac{\$13}{1} = \frac{\$455}{h}$

Solve for *h*.

$h =$ ______

Mr. Ramirez would have to work

______ hours.

2. The distance a spring stretches varies directly with the force applied to it. If a 4-pound weight stretches a spring a distance of 26.8 inches, how far will the spring stretch if a 10-pound weight is applied?

Pounds	1	2	3	4	10
Inches	?	?	?	26.8	?

Find how far the spring will stretch when a 1-pound weight is applied.

$\frac{26.8}{4} =$ ______

Complete the table.

Pounds	1	2	3	4	10
Inches					?

Find how far the spring will stretch when a 10-pound weight is applied.

$\frac{6.7}{1} = \frac{d}{10}$

Solve for *d*.

$d =$ ______

The spring will stretch ______ inches.

Name ______________________ Date ____________ Class ____________

STANDARD **7AF4.2**

Focus on CAHSEE Standards: Intervention

Sample Problems

Grade 7 • Algebra and Functions 4.2

Solve multistep problems involving rate, average speed, distance, and time or direct variation.

1. Mr. Thompson's car can go an average of 36 miles per gallon of gas. He plans on driving 288 miles today and 216 miles tomorrow. How many gallons of gas does he expect to use?

What information do you know?

____

____

What do you need to find?

____

____

What do you need to find first?

____

____

How many miles does Mr. Thompson plan to drive altogether?

____

How can you find how many gallons of gas Mr. Thompson will use?

____

____

$504 \div 36 =$ ____________

Mr. Thompson will use 14 gallons of gas.

Check your answer.

$36 \times 14 =$ ______

2. Ms. Edwards cycled 12 miles per hour for 3 hours and then 8 miles per hour for 2 hours. How many miles did Ms. Edwards cycle in all?

What information do you know?

____

____

What do you need to find?

____

What equation can you write?

Miles = (____ × ____) + (____ × ____)

How many miles will Ms. Edwards cycle in all? ____________

3. Four nights at a hotel cost $568. At that rate how much money would 7 nights cost?

The cost can be found by finding the unit price. What is the price per night?

____

Multiply the cost per night by the number of nights. ____________________

How much money does 7 nights at the hotel cost? ______

Name ______________________ Date __________ Class __________

Focus on CAHSEE Standards: Intervention

STANDARD 7AF4.2

Skills Practice

Grade 7 • Algebra and Functions 4.2

Solve multistep problems involving rate, average speed, distance, and time or direct variation.

Find the unit price to the nearest cent.

1. 32 ounces of juice for $3.84

 ____

2. $49 for a dozen roses

 ____

3. $10.68 for a case of 12 rolls of paper towels

 ____

4. a 5 pound box of clementines for $6.99

 ____

Use the unit prices in Exercises 1–4 to find the following.

5. the price of 35 ounces of juice

 ____

6. the price of 8 roses

 ____

7. the price of 15 rolls of paper towels

 ____

8. the price of 6 pounds of clementines

 ____

For Exercises 9–12, find the speed in mi/h to the nearest mile. Then answer the question.

9. A car goes 441 miles in 7 hours. How far can it go in 9 hours?

 ____

10. A bat flies 25 miles in 28 minutes. How far can it fly in an hour?

 ____

11. A plane flies 1,575 miles in $2\frac{1}{2}$ hours. How far can it fly in 6 hours?

 ____

12. A man walks 12 miles in 8 hours. How far does he walk in 3 hours?

 ____

Find the gas mileage in mi/gal. and answer the questions.

13. Joe drove 350 miles on 10 gallons of gas. How far could he drive on 12 gallons?

 ____

14. Adele drove 4 miles on $\frac{1}{4}$ gallon of gas. How far could she drive on 16 gallons?

 ____

Name ______________________ Date __________ Class __________

STANDARD 7AF4.2 Focus on CAHSEE Standards: Intervention

Problem Solving Practice

Grade 7 • Algebra and Functions 4.2

Solve multistep problems involving rate, average speed, distance, and time or direct variation.

1. Mr. Samuelson bought 3 meters of wire that have a mass of 450 grams. At the same rate, what is the mass of 10 meters of wire?

Step 1: What are you asked to find?

____

Step 2: What are the facts?
What is the mass of 3 meters of wire?

____

Step 3: Break down the problem.
Find the unit rate. (mass per meter)
Then multiply the unit rate by 10.

Step 4: Solve the problem.
Divide 450 by 3 to find the unit rate.

____

Multiply the unit rate by 10 to find the mass.

____

The mass is __________.

Step 5: Check that your answer makes sense.

$$\frac{450 \text{ grams}}{3 \text{ meters}} = \frac{\_\_\_ \text{ grams}}{10 \text{ meters}}$$

Are the ratios equivalent?

____

____

2. A restaurant owner bought a 5-pound bag of flour for $8.45. Last week the owner bought another size bag of flour at the same unit price for $40.56. How many pounds did the bag weigh?

Step 1: What are you asked to find?

____

____

Step 2: What are the facts?
How much did this week's bag of flour weigh

____

How much did this week's bag of flour cost?

____

How much did last week's bag of flour cost?

____

Step 3: Break down the problem.
Find the unit price per pound. Then divide the cost of the larger bag by the unit price to find the weight.

Step 4: Solve the problem.
Divide 8.45 by 5 to find the unit price.

____

Divide the cost of the larger bag by the unit price. __________

Step 5: Check that your answer makes sense.

____

____

Name ______________________ Date __________ Class __________

Focus on CAHSEE Standards: Intervention

STANDARD 7MG1.1

Sample Problems

Grade 7 • Measurement and Geometry 1.1

Compare weights, capacities, geometric measures, times, and temperatures within and between measurement systems (e.g. miles per hour and feet per second, cubic inches to cubic centimeters).

1. Due to gravity, objects weigh different amounts on Earth and Mars. For example, a person who weighs 90 pounds on Earth will weigh approximately 34.2 pounds on Mars. If Lucas weighs 122 pounds on Earth what is his weight on Mars?

Write a proportion using the given information. Let x be the unknown number.

$\frac{\text{90 Earth pounds}}{\text{34.2 Mars pounds}}$

= ____________

Write the proportion: $\frac{90}{34.2}$ = ________

Cross multiply: 90 • x = ________

Multiply both sides: 90x = ________

Divide each side by 90 to isolate the variable: $\frac{90x}{90}$ = ________

Solve: x = ________

Round to the nearest tenth: ________

Approximately how much does Lucas weigh on Mars?

____

2. The Golden Gate Bridge is approximately 1.7 miles long. About how many feet is this?

Use a unit conversion for miles to feet. Let x be the unknown length.

$\frac{\text{5,280 feet}}{\text{1 mile}} = \frac{x}{\text{1.7 miles}}$

Cross multiply: ________

Multiply both sides: ________

Approximately how long is the bridge?

____

3. Former pitcher Nolan Ryan once threw a baseball 100.8 miles per hour. Approximately how many feet per second was Ryan's pitch?

Simplify the conversion to express the speed of the ball in feet per second.

$\frac{\text{60 mph}}{\text{90 feet per second}} = \frac{60}{90} = \frac{6}{9} = \frac{2}{3}$

Let x be the unknown distance:

Write the proportion:

$\frac{\text{2 mph}}{\text{3 feet per second}} = \frac{\text{100.8 mph}}{x \text{ feet per second}}$

Cross multiply: ________

Multiply both sides: ________

Isolate the variable: ________

Approximately how many feet per second was Ryan's pitch?

____

Name ______________________ Date ____________ Class ____________

STANDARD **7MG1.1**

Focus on CAHSEE Standards: Intervention

Skills Practice

Grade 7 • Measurement and Geometry 1.1

Compare weights, capacities, geometric measures, times, and temperatures within and between measurement systems (e.g. miles per hour and feet per second, cubic inches to cubic centimeters).

Show your work.

1. Jolene's bowling ball weighs 8 pounds. How many ounces does it weigh?

2. An elephant weighs 12.5 tons, or 25,000 pounds. A blue whale weighs 150 tons. How many pounds does the whale weigh?

3. At a gas station in Canada, Gill buys 34 liters of gas, or about 9 gallons. At his next fill-up, he buys 26 liters. How many gallons of gas is 26 liters? (Round the answer to the nearest tenth.)

4. Including the end zone, a regulation football field is 360 feet long by 160 feet wide. How many yards wide is a football field?

5. Gert biked 22 miles from her home to her grandmother's home. She made the trip in 2.5 hours. What was her average speed in miles per hour?

Name ______________________ Date ____________ Class ____________

STANDARD 7MG1.2

Focus on CAHSEE Standards: Intervention

Sample Problems

Grade 7 • Measurement and Geometry 1.2

Construct and read drawings and models made to scale.

1. In preparation for a road trip, Kendra looked at a map whose scale is 1 inch : 75 miles. On the map, the distance she wants to cover is 2.4 inches long. How many miles long is the actual trip?

   Let *d* represent the actual distance. Complete a proportion using the scale and the given map distance.

   $\frac{\text{actual distance}}{\text{map distance}} = \frac{\square}{1} = \frac{\square}{2.4}$

   Cross multiply to solve the proportion for *d*.

   ____ d = ____ • ____

   d = ______

   How many miles is the actual trip?

   ____

2. Darryl has a photograph that is 6 inches long and 4 inches wide. He wants to make a copy of it that is 9 inches long and 6 inches wide. By what scale factor should he enlarge the photograph?

   Compare the length of the larger photo to the length of the smaller one.

   $\frac{\text{length in larger photo}}{\text{length in smaller photo}}$ = ____

   Write as a decimal or percent.

   ____ = ____ = ______%

   By what scale factor should he enlarge the photograph? ____

3. Bea has a photograph that is 60 inches long and 40 inches wide. She wants to reduce the photograph by 40%. What will be the new dimensions?

   Find 40% of each dimension.

   40% of 60 = ____

   40% of 40 = ____

   The new photograph will be ____ inches long and ____ inches wide.

4. The two right-triangle gardens shown are similar. How long is the side of the garden whose endpoints are *X* and *Z*?

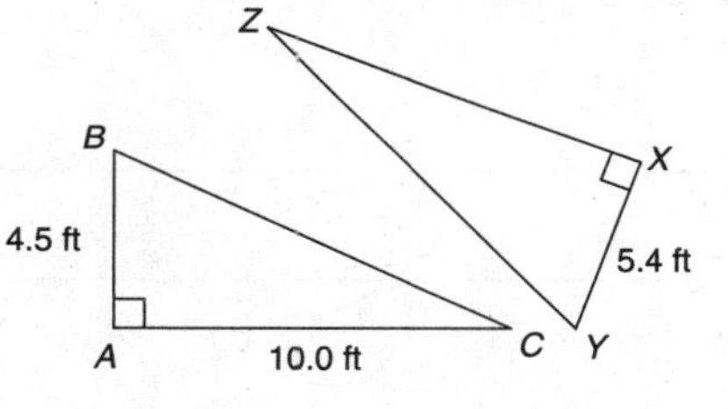

   $\overline{AB} \longleftrightarrow$ ________

   $\overline{AC} \longleftrightarrow$ ________

   Write a proportion using known measurements and *x* for $\overline{XZ}$.

   $\frac{\text{smaller triangle}}{\text{larger triangle}} \quad \frac{AB}{XY} = \frac{\square}{\square}$

   $\frac{4.5}{5.4} = \frac{\square}{\square}$

   Solve the proportion for *x*. x = ______.

Name ______________________ Date __________ Class __________

STANDARD 7MG1.2

Focus on CAHSEE Standards: Intervention

Skills Practice

Grade 7 • Measurement and Geometry 1.2

Construct and read drawings and models made to scale.

Show your work.

1. A scale model of a car is 10 inches long. The scale factor is 1 : 18. How many feet long is the actual car?

____

____

____

2. Zeb is planning to remodel a rectangular room. The room's dimensions are 15 feet by 12 feet. Zeb makes a blueprint with a scale factor of 1 inch : 2.5 feet. How long is the longer dimension of the room in the blue print?

____

____

____

____

____

3. 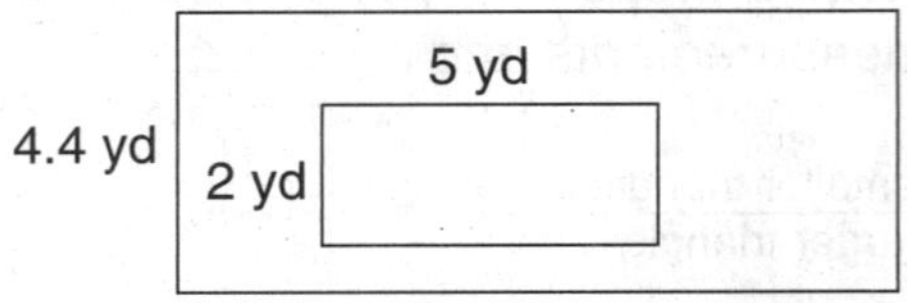

The two rectangular gardens shown are similar. What is the length of the larger garden?

____

____

____

____

____

____

Answer the questions.

4. A certain map uses a scale of 1 centimeter for every 25 kilometers. How many kilometers is 5 centimeters on the map? ______

5. Anita made a scale sculpture of a skateboard. She used a scale factor of 1 : 20. The sculpture is 45 feet long. How long is the original skateboard?

____

6. An architect made a scale model of a skyscraper. The model uses a scale factor of 1 : 500. The model is 2 feet tall. How tall will the completed skyscraper be? __________

7. Liz bought a scale model of the Eiffel Tower for her little sister. It is 2 feet tall. The Eiffel Tower is 990 feet tall. What is the scale factor? __________

8. Maia bought a scale model of the Washington Monument for her mother. It is 6 inches tall. The Washington Monument is 555 feet tall. What is the scale factor? __________

Name ______________________ Date ____________ Class ____________

STANDARD 7MG1.3

Focus on CAHSEE Standards: Intervention

Alternative Strategies

Grade 7 • Measurement and Geometry 1.3

Use measures expressed as rates (e.g. speed, density) and measures expressed as products (e.g. person-days) to solve problems; check the units of the solutions; and use dimensional analysis to check the reasonableness of the answer.

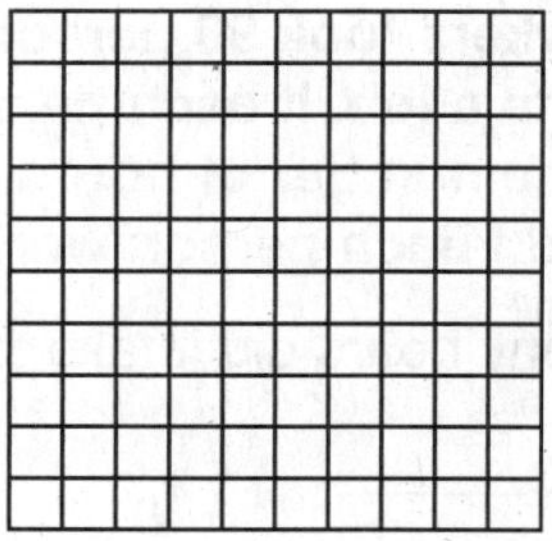

Use the grid.

A cyclist traveled 57 miles in 3 hours. At this rate, how far would the cyclist travel in 5 hours?

1a. Shade 57 squares of grid paper.

1b. Cut the paper into 3 sections so that each section has an equal number of shaded squares. Each section represents 1 hour.

1c. How many squares are in one section?

______

1d. What is the cyclist's rate of speed?

____

1e. Shade two more sections of 19 squares each.

1f. Count all the shaded squares. How far would the cyclist travel in 5 hours?

____

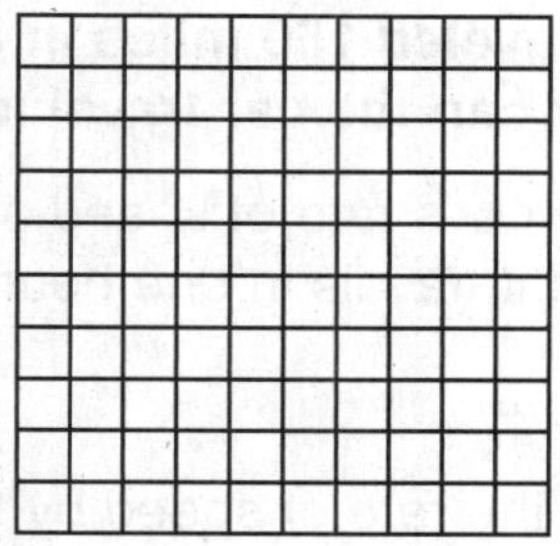

Use the grid.

It took 6 carpenters 45 person-hours to complete a job. If each carpenter worked the same number of hours, how long did each person work?

2a. Shade 45 squares of grid paper.

2b. Cut the paper into 6 sections so that each section has an equal number of whole shaded squares. Each section represents 1 hour.

2c. How many whole squares are in one section? ______

2d. How many squares are left? ______

2e. How can these single squares be divided into 6 equal pieces?

____

2f. How many shaded squares make up one equal part? ______

2g. How many person-hours did each carpenter work? ______

Name ______________________ Date ____________ Class ____________

STANDARD 7MG1.3

Focus on CAHSEE Standards: Intervention

Sample Problems

Grade 7 • Measurement and Geometry 1.3

Use measures expressed as rates (e.g. speed, density) and measures expressed as products (e.g. person-days) to solve problems; check the units of the solutions; and use dimensional analysis to check the reasonableness of the answer.

1. **A car traveled 150 miles in 3 hours. How far can the car travel in 7 hours?**

   Find the car's rate of speed or average distance it travels in one hour.

   $150 \div 3 =$ ______

   Multiply the rate of speed by the amount of time the car will travel.

   $50 \times 7 =$ ______

   How far can the car travel in 7 hours?

   ____

2. **The mass of a rock sample is 51 g. The volume of the sample is 17 cm^3. What is the density of the sample?**

   Density is mass per unit of volume.

   What division sentence can you use to find the mass of one gram?

   ______ □ ______

   Divide. ______________

   What is the density of the sample?

   ____

3. **Ten workers took 90 person-hours to complete a job. If each person worked the same number of hours, how many hours did each person work?**

   How many hours did it take to complete the job? ______

   How many workers were there? ______

   What division sentence can you use to find how many hours one person worked?

   ______ □ ______

   Divide. ______________

   How many hours did each person work?

   ____

4. **It took 8 workers 6 person-days each to complete a job. What was the total number of person-days needed to complete the job?**

   How many workers were there? ______

   How many person-days did one worker work? ______

   What multiplication sentence can you use to find the total number of person-days worked?

   ______ □ ______

   How many days did it take to complete the job? ______________

Name ______________________ Date ____________ Class ____________

STANDARD 7MG1.3

Focus on CAHSEE Standards: Intervention

Skills Practice

Grade 7 • Measurement and Geometry 1.3

Use measures expressed as rates (e.g. speed, density) and measures expressed as products (e.g. person-days) to solve problems; check the units of the solutions; and use dimensional analysis to check the reasonableness of the answer.

1. A cyclist traveled 28.6 miles in 2 hours. What was the cyclist's rate of speed?

____

2. A car traveled 100.5 miles in 3 hours. What was the car's rate of speed?

____

3. The mass of a rock sample is 96 g. The volume of the sample is 16 cm^3. What is the density of the sample?

____

4. A cyclist traveled 60.6 miles in 3 hours. How far would the cyclist travel in 9 hours?

____

5. A turtle walked $\frac{1}{10}$ mile in 20 minutes. How far will the turtle walk in 1 hour?

____

For questions 6–8, find the person hours for each worker.

6. 14 workers took 35 person-hours to complete a job.

____

7. 21 workers took 88.2 hours to complete a job.

____

8. 13 workers took 48.1 hours to complete a job.

____

For questions 9–10, find the total person-days worked.

9. 12 workers took 11 person-days each to complete a job.

____

10. 20 workers took 5.5 person-days each to complete a job.

____

Name ______________________ Date __________ Class __________

STANDARD 7MG1.3

Focus on CAHSEE Standards: Intervention

Problem Solving Practice

Grade 7 • Measurement and Geometry 1.3

Use measures expressed as rates (e.g. speed, density) and measures expressed as products (e.g. person-days) to solve problems; check the units of the solutions; and use dimensional analysis to check the reasonableness of the answer.

1. The chart below describes the speed of four drivers.

Driver	Distance	Time
Alba	177 miles	3 hours
Chloe	274.5 miles	4.5 hours
Marco	348 miles	6 hours
Zack	150 miles	2.5 hours

Which driver has the *fastest* rate of speed?

Step 1: What question do you need to answer?

____

____

Step 2: What do you know?

____

____

Step 4: Which driver had the fastest rate of speed?

____

____

____

____

____

____

Step 5: How can you check your calculations for each driver?

____

____

2. Six painters worked a total of 27 person-days to paint a building. How many person-days did each painter work?

Step 1: What question do you need to answer?

____

____

Step 2: What do you know?

____

____

Step 3: What division sentence can you use to find how long each painter worked?

____

Step 4: How long did each painter work?

____

Step 5: How can you check your answer?

____

____

____

____

Name ______________________ Date ____________ Class ____________

STANDARD 7MG2.1

Focus on CAHSEE Standards: Intervention

Sample Problems

Grade 7 • Measurement and Geometry 2.1

Use formulas routinely for finding the perimeter and area of basic two-dimensional figures and the surface area and volume of basic three-dimensional figures, including rectangles, parallelograms, trapezoids, squares, triangles, circles, prisms, and cylinders.

1. What is the area of trapezoid *JKLM* in square units? ____________

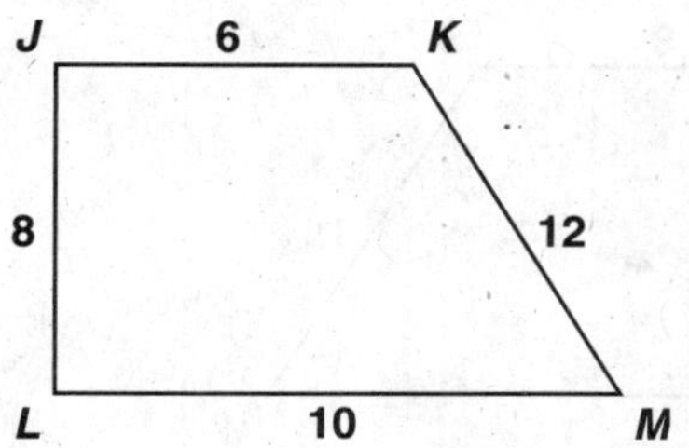

What are the lengths of b_1 and b_2?

____

What is the sum of $b_1 + b_2$? ____

What is the height of the trapezoid? ____

What is $\frac{1}{2}$ of 8? ____

What is the product of 4×16? ____

What is the area of the trapezoid?

____

2. One side of a regular pentagon is 8.2 cm long. What is the perimeter of the pentagon?

How many sides does a pentagon have?

____

What multiplication sentence can you use to find the perimeter? ____________

What is the product? ____

What is perimeter of the pentagon?

____

3. What is the volume of a cylinder with a diameter of 8 cm and a height of 12 cm? ____________

What is the radius of the cylinder? ____

What is r^2? ____

What is the height? ________

Write the formula for volume of the cylinder, replacing the variables with the values named. Use $\pi = 3.14$.

____

Multiply to find the volume of the cylinder. ____________

4. A gift box is 8 in. long, 5 in. wide, and 3 in. high. How much paper is needed to wrap the box?

____

What is the product of the box's length and height? ____________

What is the product of the box's length and width? ____________

What is the product of the box's width and height? ____________

Rewrite the formula, replacing the variables with the values named.

____

How much paper is needed to wrap the box? ________

Name ______________________ Date ____________ Class ____________

Focus on CAHSEE Standards: Intervention

STANDARD 7MG2.1

Skills Practice

Grade 7 • Measurement and Geometry 2.1

Use formulas routinely for finding the perimeter and area of basic two-dimensional figures and the surface area and volume of basic three-dimensional figures, including rectangles, parallelograms, trapezoids, squares, triangles, circles, prisms, and cylinders.

1. A can of soup has a height of 6 inches and a diameter of 4 inches. What is the volume of the can?

Use 3.14 as an estimate for π. ($V = \pi r^2 h$)

____

2. A circular flower garden has a diameter of 8 feet. What is the area of the garden? Use 3.14 as an estimate for π. ($A = \pi r^2$)

____

3. How much wrapping paper is needed to cover the gift box shown below?

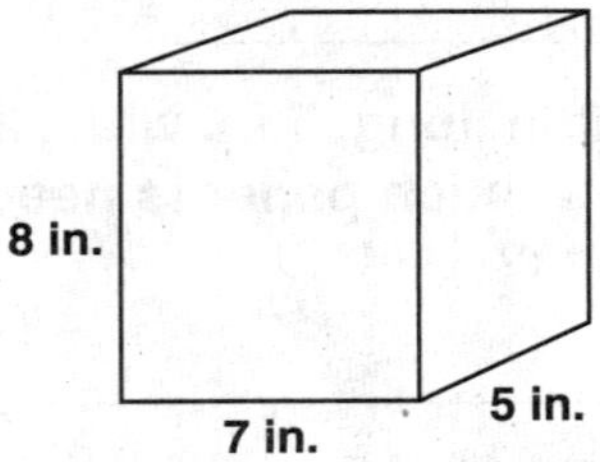

____

4. One side of a regular hexagon is 5 cm long. What is the perimeter of the hexagon?

____

5. What is the area of trapezoid *RSTW* in square units? ________________.

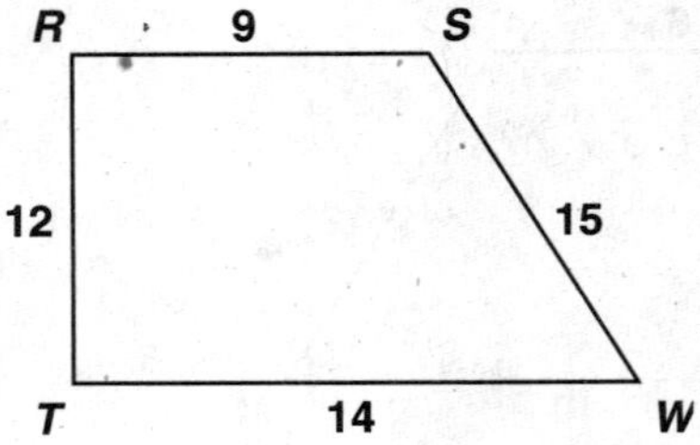

____

6. What is the area of triangle *ABC*?

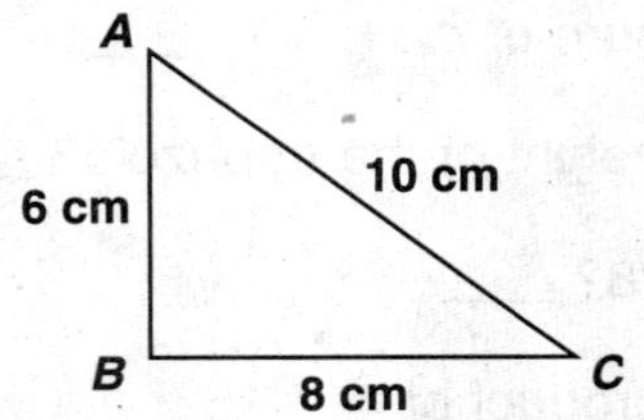

____

7. What is the volume of a rectangular prism that is 4 cm long, 5 cm high, and 6 cm wide?

____

8. Dixie wants to put lace trim around the edges of a square quilt. The area of the quilt is 36 ft^2. How much lace does Dixie need for the trim?

____

Name ______________________ Date __________ Class __________

STANDARD 7MG2.2

Focus on CAHSEE Standards: Intervention

Sample Problems

Grade 7 • Measurement and Geometry 2.2

Estimate and compute the area of more complex or irregular two- and three-dimensional figures by breaking the figures down into more basic geometric objects.

1. Mitch cut four congruent triangles off the corners of a rectangle to make an octagon, as shown below.

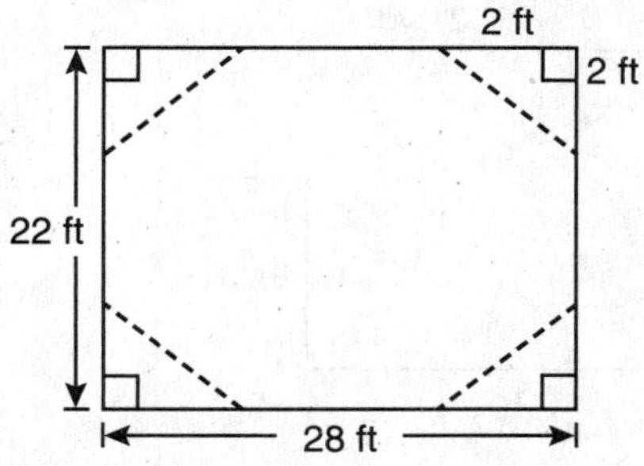

What is the area of the octagon?

What is the area of the entire rectangle?

____

What is the area of one triangle? ________

What is the area of all four triangles?

____

Subtract the area of the four triangles from the area of the entire rectangle.

What is the difference? ________

What is the area of the octagon?

____

2. What is the area of this irregular figure?

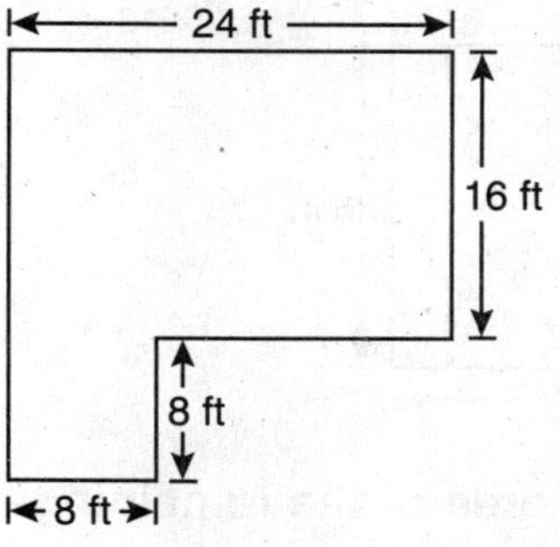

Divide the figure into a rectangle and a square. Draw a line across the bottom of the figure so that a 8 ft by 8 ft square forms. What is the length and width of the remaining rectangle? ________

What is the area of the 8 ft by 8 ft square? ________

What is the area of the rectangle?

____

What is the sum of the areas? ________

What is the area of the irregular figure?

____

Name ______________________ Date ____________ Class ____________

STANDARD **7MG2.2**

Focus on CAHSEE Standards: Intervention

Skills Practice

Grade 7 • Measurement and Geometry 2.2

Estimate and compute the area of more complex or irregular two- and three-dimensional figures by breaking the figures down into more basic geometric objects.

1. Rita cut four squares from the corners of a piece of fabric.

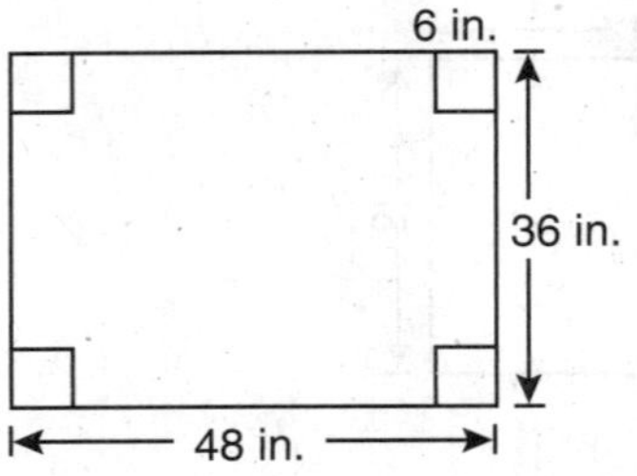

What is the area of the remaining fabric?

____

2. What is the area of the irregular figure shown below?

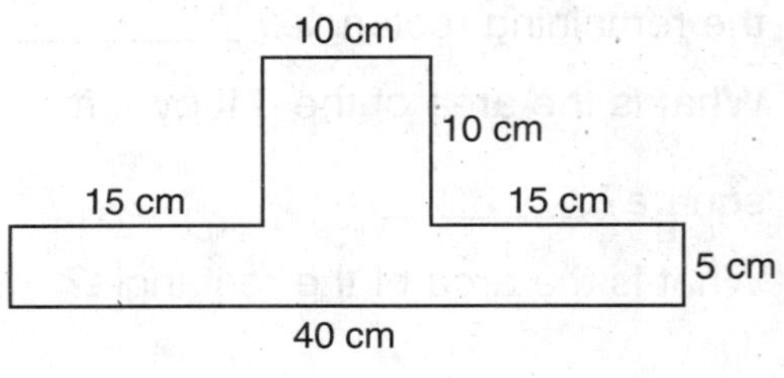

____

3. What is the area of the white walkway around the shaded rectangular garden?

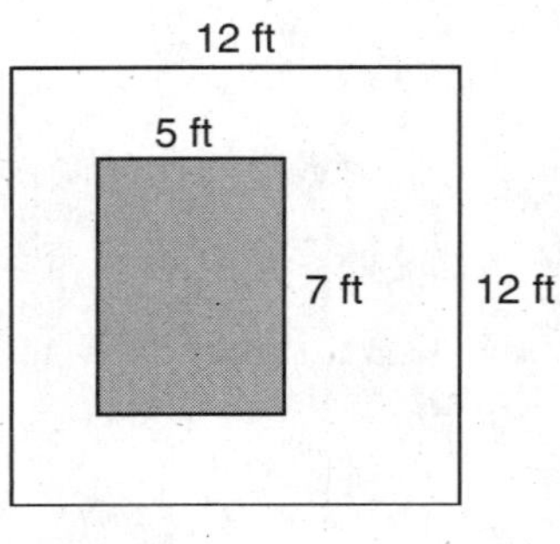

____

4. The diagram shows the dimensions of a room. What is the area of the room?

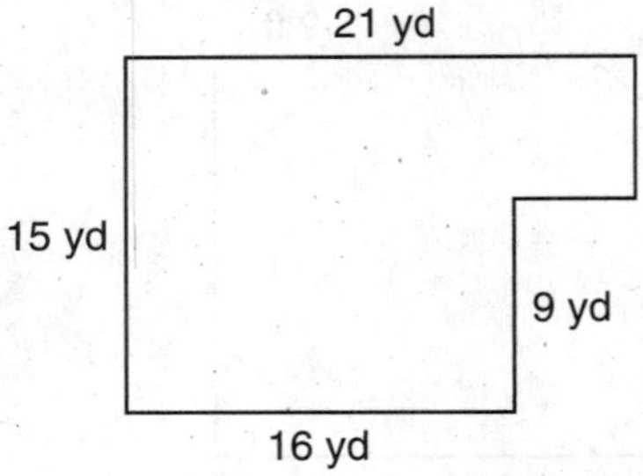

____

5. Hillary cut congruent triangles from the corners of a rectangle. What is the area of the octagon that formed?

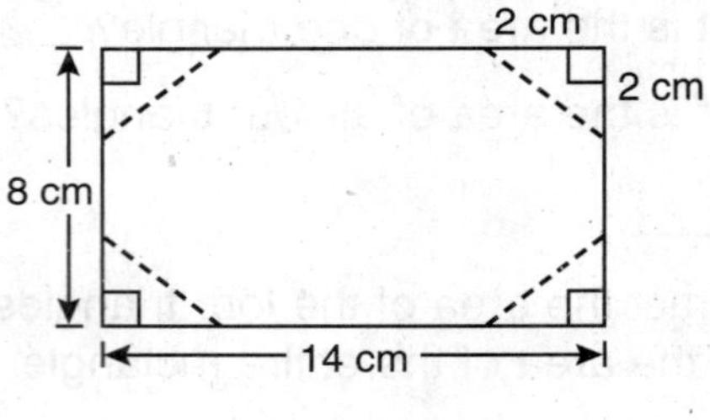

____

6. Yiang made a sketch of a wall she is painting. The shaded areas show two windows. What is the area of the wall to be painted?

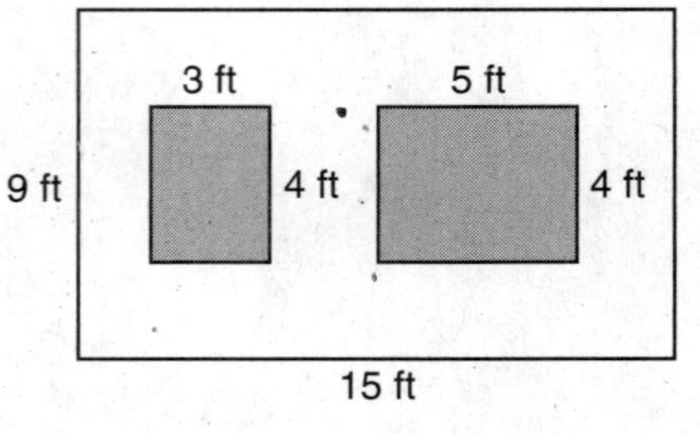

____

Name ______________________ Date __________ Class __________

STANDARD 7MG2.3

Focus on CAHSEE Standards: Intervention

Sample Problems

Grade 7 • Measurement and Geometry 2.3

Compute the length of the perimeter, the surface area of the faces, and the volume of a three-dimensional object built from rectangular solids. Understand that when the lengths of all dimensions are multiplied by a scale factor, the surface area is multiplied by the square of the scale factor and the volume is multiplied by the cube of the scale factor.

1. A box of sugar cubes is 15 cm tall, 12 cm long, and 8 cm wide. What is the maximum number of 1-cm sugar cubes that can fit inside the box?

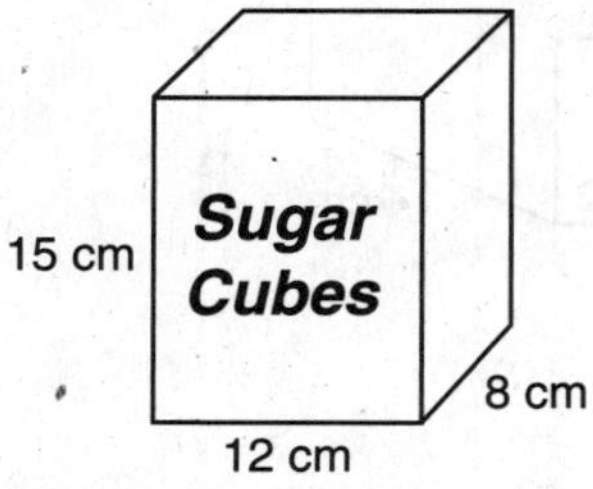

What question do you need to answer? ______________________

____

What do you know? ______________________

____

____

What formula will you use to calculate the volume of the box?

____

What is the volume of the box?

____

How many 1-cm cubes can fit inside the box? __________

2. A cereal box measures 10 in. tall, by 6 in. long, by 2.5 in. wide. How much cardboard is needed to make the box?

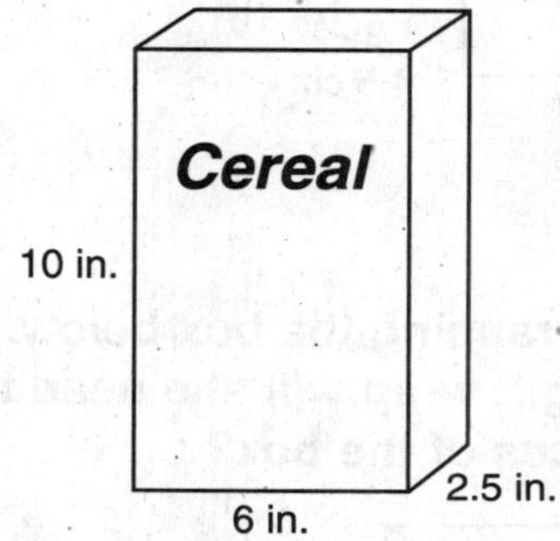

What question do you need to answer? ______________________

____

What do you know? ______________________

____

____

What formula will you use to calculate the surface area of the box?

____

____

What is the surface area of the box?

____

How much cardboard is needed to make the box? __________

Name _______________ Date _______ Class _______

STANDARD 7MG2.3 Focus on CAHSEE Standards: Intervention
Skills Practice

Grade 7 • Measurement and Geometry 2.3

Compute the length of the perimeter, the surface area of the faces, and the volume of a three-dimensional object built from rectangular solids. Understand that when the lengths of all dimensions are multiplied by a scale factor, the surface area is multiplied by the square of the scale factor and the volume is multiplied by the cube of the scale factor.

1. What is the volume of the rectangular solid shown below?

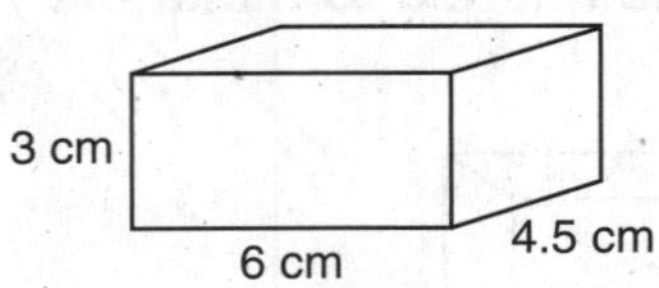

____

2. Yasmin is wrapping the box below. How much gift wrap will she need to cover all faces of the box?

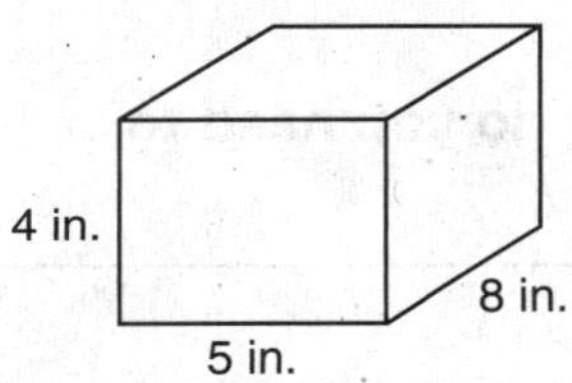

____

3. Melba made a square banner 4 feet on a side. She sewed a 6 in. border to all sides of the banner. What is the perimeter of Melba's banner?

____

4. The dimensions of this rectangular solid are doubled. What is the volume of the new solid?

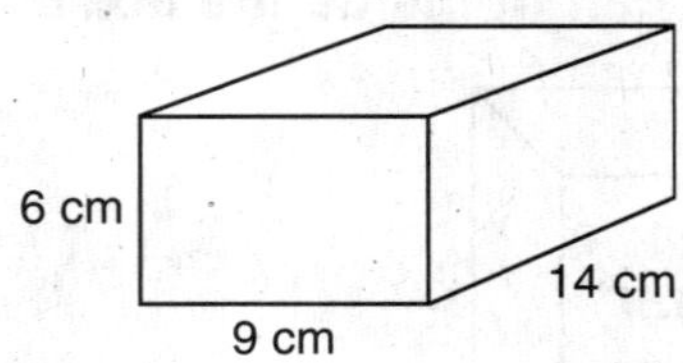

____

5. How much cardboard was needed to make the carton below?

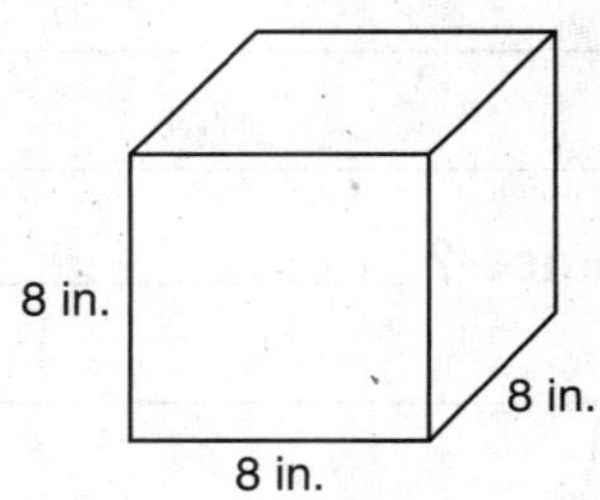

____

6. If all sides of the carton described in Problem 5 are cut in half, what is the volume of the smaller carton?

____

Name ______________________ Date __________ Class __________

Focus on CAHSEE Standards: Intervention

STANDARD 7MG2.4

Sample Problems

Grade 7 • Measurement and Geometry 2.4

Relate changes in measurement with a change of scale to the units used (e.g., square inches, cubic feet) and to conversions between units (1 square foot = 144 square inches or [1 ft^2 = 144 in^2], 1 cubic inch is approximately 16.38 cubic centimeters or [1 in^3 = 16.38 cm^3]).

1. What is the volume of the carton shown below expressed in cubic centimeters? ______________

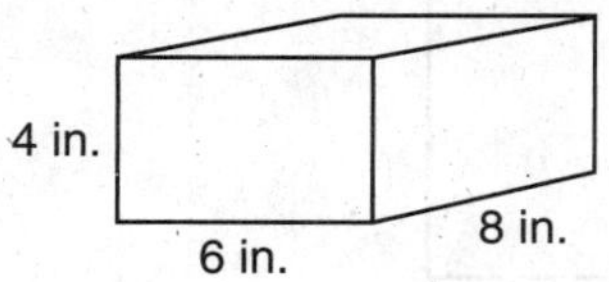

What question do you need to answer?

____

____

What do you know? ______________

____

____

What formula will you use to calculate the volume of the box?

____

What is the approximate volume of the box expressed in cubic inches?

____

How will you express this volume in cubic centimeters?

____

About how many cubic centimeters will fit in the box?

____

2. The diagram shows the dimensions of Troy's living room. The carpeting is sold by the square yard. How many square yards does Troy need?

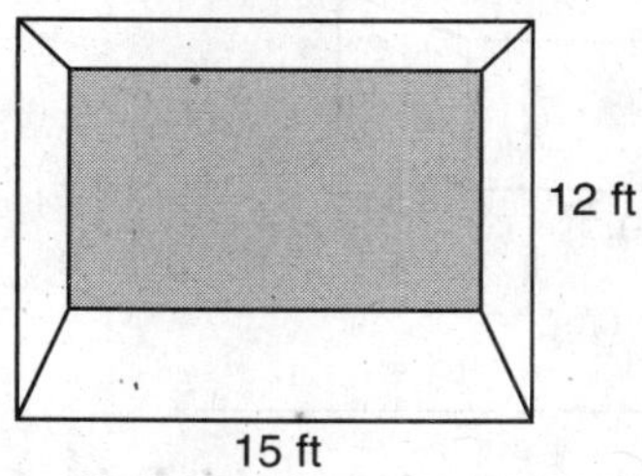

What question do you need to answer?

____

____

What do you know? ______________

____

What formula will you use to find the area of the room? ______________

What is the area of the room?

____

How will you express this area in square yards? ______________

How many square yards does Troy need? ______________

Name ______________________ Date __________ Class __________

STANDARD 7MG2.4

Focus on CAHSEE Standards: Intervention

Skills Practice

Grade 7 • Measurement and Geometry 2.4

Relate changes in measurement with a change of scale to the units used (e.g., square inches, cubic feet) and to conversions between units (1 square foot = 144 square inches or 1 ft^2 = 144 in^2, 1 cubic inch is approximately 16.38 cubic centimeters or 1 in^3 = 16.38 cm^3).

1. What is the volume of the carton shown below expressed in cubic centimeters? __________

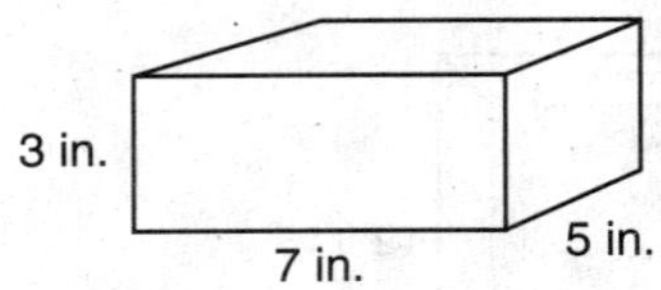

____

2. Dee has a square piece of fabric. One side of the fabric measures 66 inches. What is the area of the fabric expressed in square feet?

____

3. Rick is buying carpeting for a room that is 14 ft long and 12 ft wide. How many square yards of carpet does he need?

____

4. The volume of a box is 272 cm^3. What is the volume of the box expressed in cubic inches?

____

5. How many square yards of flooring are needed to cover the floor of this room?

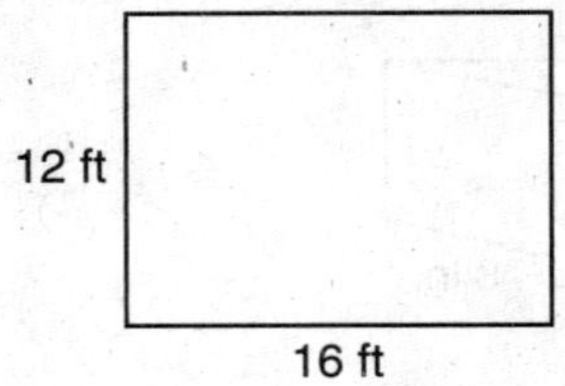

____

6. Ella made a quilt that is 4 ft long and 6 ft wide. Tess made a quilt with dimensions that are double Ella's quilt. What is the area of Tess's quilt given in square yards?

____

7. What is the volume of the rectangular solid shown below expressed in cubic centimeters?

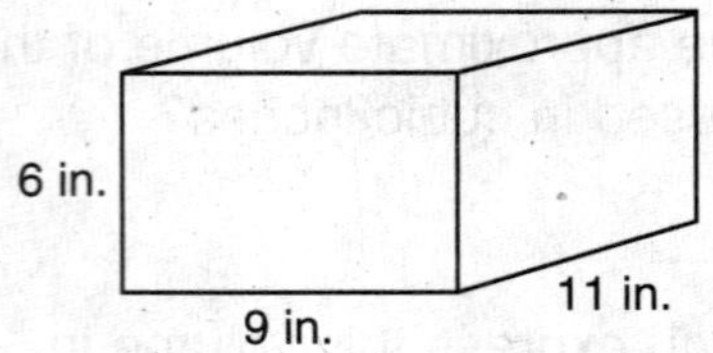

____

Name ______________________________ Date __________ Class __________

STANDARD **7MG3.2**

Focus on CAHSEE Standards: Intervention

Sample Problems

Grade 7 • Measurement and Geometry 3.2

Understand and use coordinate graphs to plot simple figures, determine lengths and areas related to them, and determine their image under translations and reflections.

The vertices of triangle *ABC* are *A*(−2, −2), *B*(3,5) and *C*(4, −2).

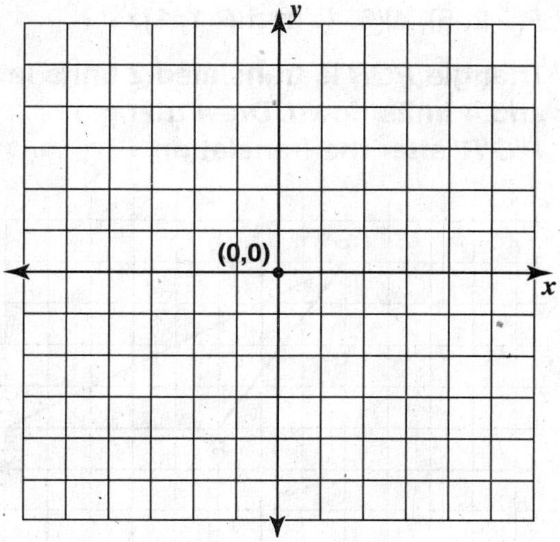

1. Draw triangle *ABC* on the grid.

Plot vertices *A, B,* and *C.* Start at the origin (0,0). Point *A*(−2, −2) is left 2 units and down 2 units. Start at the origin (0,0). Point *B*(3,5) is right 3 units and up ______ units. Start at the origin (0,0). Point *C*(4, −2) is right 4 units and down ______ units. Connect the vertices to show the triangle.

2. What is the length of $\overline{AC}$?

Since $\overline{AC}$ is parallel to the *x*-axis, the length of $\overline{AC}$ is the difference in the *x* coordinates,

$4 - (-2) =$ ______ units.

3. Triangle *ABC* is translated right 2 units and down 3 units. Draw the image triangle *A'B'C'*. What are the coordinates?

A' ________

B' ________

C' ________

4. What is the area of triangle *ABC* above?

The formula to find the area of a triangle is the base times the height divided by 2 or $A = \frac{b \times h}{2}$.

$\overline{AC}$ is the base of the triangle.

The length of the base is ______.

The height is the length of $\overline{BD}$ drawn from point *B* to $\overline{AC}$. The length of the height is the difference in the *y*-coordinate of point *B* and the *y*-coordinate of point *D*.

$5 - (-2) =$ ______

The area of triangle *ABC* is $\frac{(6 \times \square)}{\square}$.

The area is ______ square units.

Name ______________________ Date ___________ Class ___________

Focus on CAHSEE Standards: Intervention

STANDARD 7MG3.2 *Skills Practice*

Grade 7 • Measurement and Geometry 3.2

Understand and use coordinate graphs to plot simple figures, determine lengths and areas related to them, and determine their image under translations and reflections.

1. The vertices of triangle *ABC* are *A*(−3, 2), *B*(5, 2) and *C*(−3, −3). Draw the triangle on the grid.

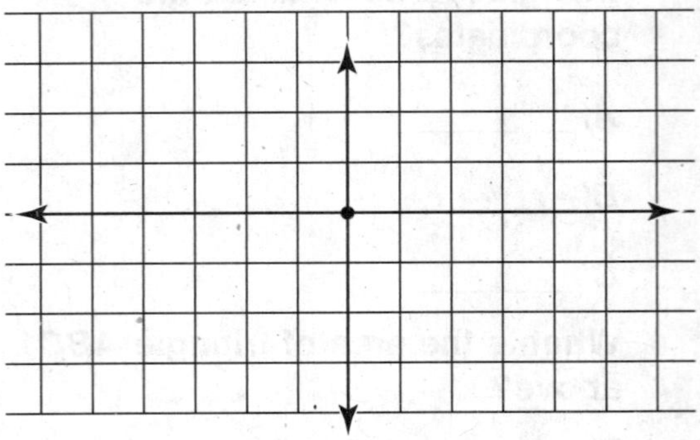

What is the length of $\overline{AB}$? __________

What is the area of triangle *ABC*?

____

2. The vertices of rectangle *DEFG* are *D*(−4, 3), *E*(5, 3), *F*(5, −2), and *G*(−4, −2).

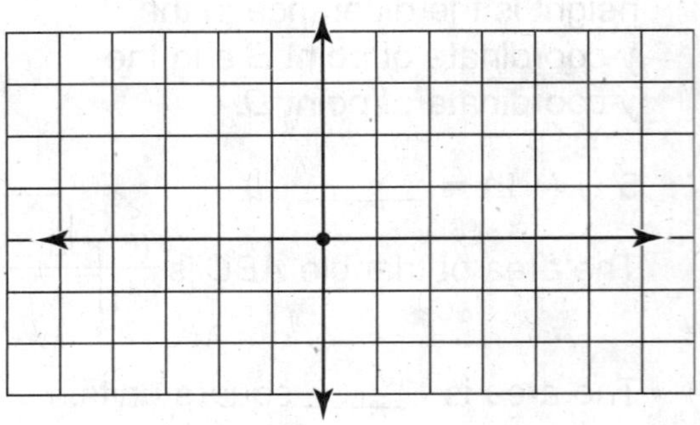

What is the length of $\overline{DE}$? __________

What is the length of $\overline{EF}$? __________

What is the perimeter of rectangle *DEFG*? __________

What is the area of rectangle *DEFG*?

____

3. The vertices of triangle *PQR* are *P*(−2, 5), *Q*(6, 1) and *R*(1, 1).

Triangle *PQR* is translated 2 units left and 4 units down. Draw triangle *P'Q'R'* after the translation.

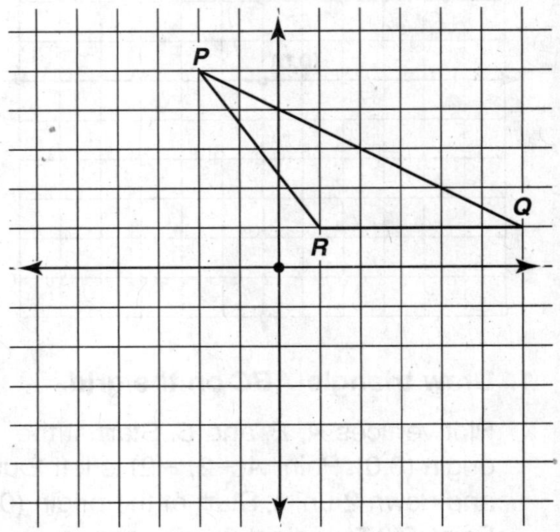

Name ______________________ Date __________ Class __________

STANDARD 7MG3.3

Focus on CAHSEE Standards: Intervention

Alternative Strategies

Grade 7 • Measurement and Geometry 3.3

Know and understand the Pythagorean theorem and its converse and use it to find the length of the missing side of a right triangle and the lengths of other line segments and, in some situations, empirically verify the Pythagorean theorem by direct measurement.

Use a centimeter ruler and the scale 0.5 cm = 1 mi or 1 cm = 2 mi to find the missing length.

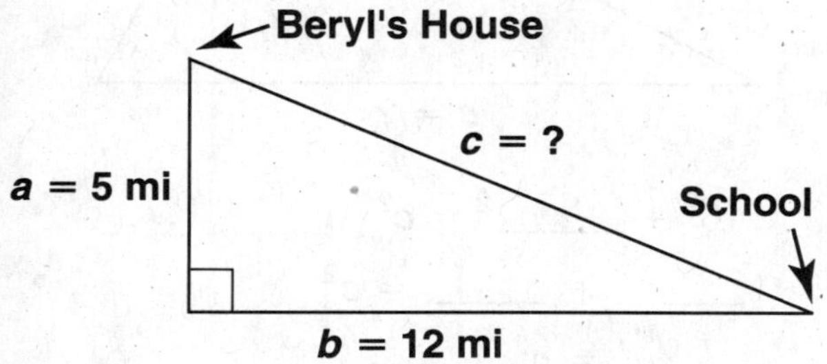

Find the distance between Beryl's house and school.

1. Distance = ______ centimeters

2. ______ cm = ______ miles

Use the Pythagorean Theorem to check.

3a. Substitute the values for *a* and *b*.

$c^2 = 5^2 +$ ______

3b. Simplify.

$c^2 =$ ______ $+ 144$

$c^2 = 169$

3c. Use the square root principle to solve for *c*.

$\sqrt{c^2} = \sqrt{169}$

$c = 13$

The distance from Beryl's house to school is ______ miles.

Use a centimeter ruler and the scale 0.5 cm = 1 mi or 1 cm = 2 mi to find the missing length.

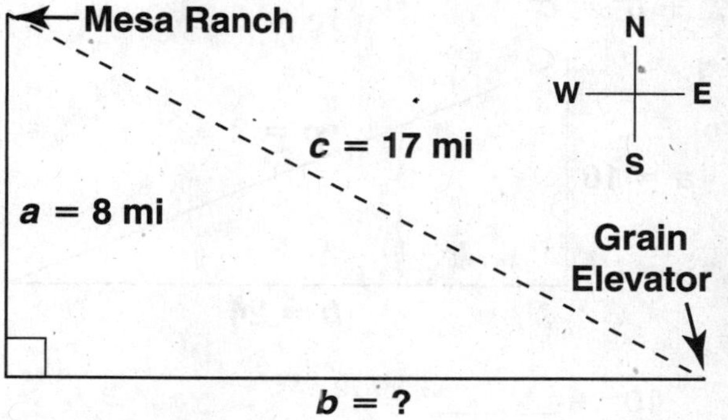

Grain is transported by truck from Mesa Ranch to the Grain Elevator. The direct route is 17 miles but there is no road. The truck must travel south 8 miles and then turn east. What is the distance, in miles, from the turn?

3. Distance = ______ centimeters

4. ______ cm = ______ miles

5. Use the Pythagorean Theorem to check.

$a^2 = b^2 + c^2$

$b^2 = c^2 - a^2$

$b = \sqrt{c^2 - a^2}$

Subtract a^2 from both sides and use the square root principle to solve for *b*.

6a. Substitute the values for *a* and *c*.

$b = \sqrt{17^2 - \_\_^2}$

6b. Simplify.

$b = \sqrt{\_\_\_\_ - 64}$

$b = \sqrt{225} =$ ______

The distance is ______ miles.

Name ______________________ Date ____________ Class ____________

Focus on CAHSEE Standards: Intervention

STANDARD 7MG3.3 *Sample Problems*

Grade 7 • Measurement and Geometry 3.3

Know and understand the Pythagorean theorem and its converse and use it to find the length of the missing side of a right triangle and the lengths of other line segments and, in some situations, empirically verify the Pythagorean theorem by direct measurement.

For questions 1–3, use the Pythagorean Theorem to find the missing side in the right triangles.

$a^2 + b^2 = c^2$

1.

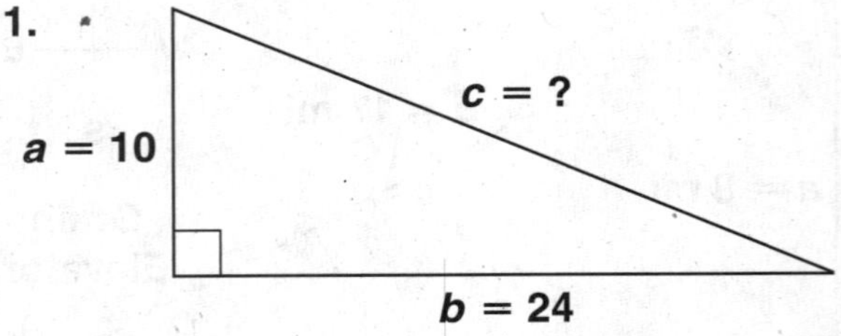

$10^2 + \underline{\qquad}^2 = c^2$

$100 + \underline{\qquad} = c^2$

$\underline{\qquad} = c^2$

$\sqrt{\underline{\qquad}} = c$

$\underline{\qquad} = c$

2.

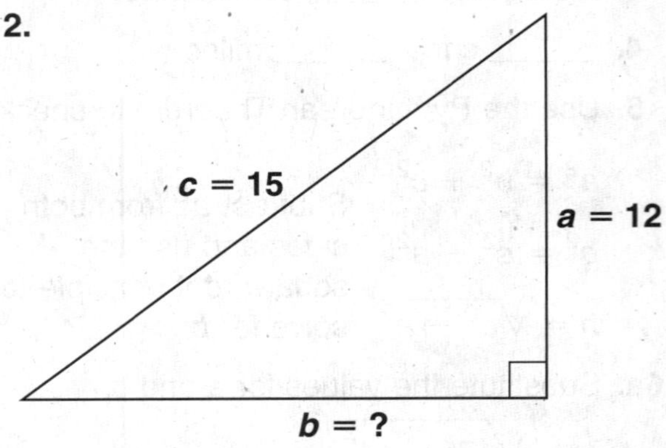

$12^2 + b^2 = \underline{\qquad}$

$144 + b^2 = \underline{\qquad}$

$b^2 = \underline{\qquad} - 144$

$b^2 = \underline{\qquad}$

$b = \underline{\qquad}$

3.

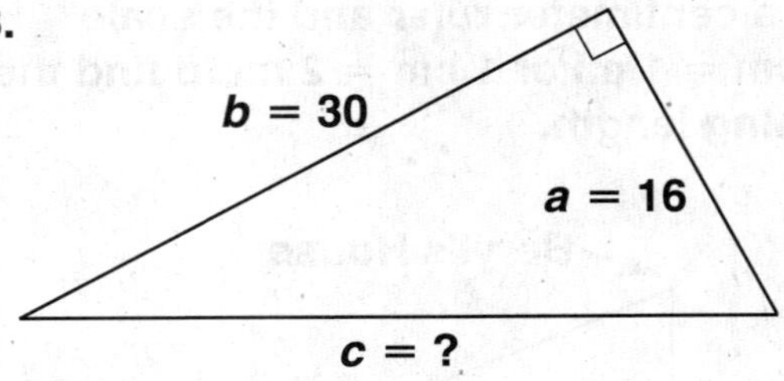

$16^2 + \underline{\qquad}^2 = c^2$

$\underline{\qquad} + \underline{\qquad} = c^2$

$\underline{\qquad} = c^2$

$\sqrt{\underline{\qquad}} = c$

$\underline{\qquad} = c$

In questions 4–5, *a* and *b* are the legs of a right triangle and *c* is the hypotenuse. Find the missing side.

4. $a = 12$, $b = 16$, $c = ?$

$\underline{\qquad}^2 + 16^2 = c^2$

$\underline{\qquad} + \underline{\qquad} = c^2$

$\underline{\qquad} = c^2$

$\underline{\qquad} = c$

5. $a = 15$, $b = ?$, $c = 39$

$\underline{\qquad}^2 + b^2 = \underline{\qquad}^2$

$b^2 = \underline{\qquad}^2 - \underline{\qquad}^2$

$b^2 = \underline{\qquad} - \underline{\qquad}$

$b^2 = \underline{\qquad}$

$b = \underline{\qquad}$

Name ______________________ Date __________ Class __________

STANDARD 7MG3.3

Focus on CAHSEE Standards: Intervention

Skills Practice

Grade 7 • Measurement and Geometry 3.3

Know and understand the Pythagorean theorem and its converse and use it to find the length of the missing side of a right triangle and the lengths of other line segments and, in some situations, empirically verify the Pythagorean theorem by direct measurement.

Use the Pythagorean theorem to find the missing side in the right triangles.

$a^2 + b^2 = c^2$

1.

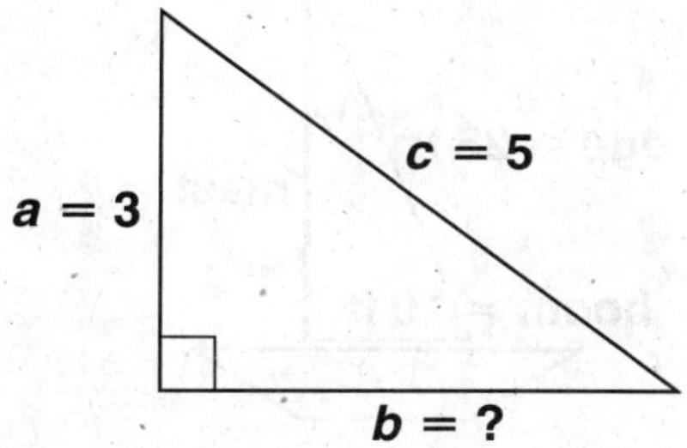

____

2.

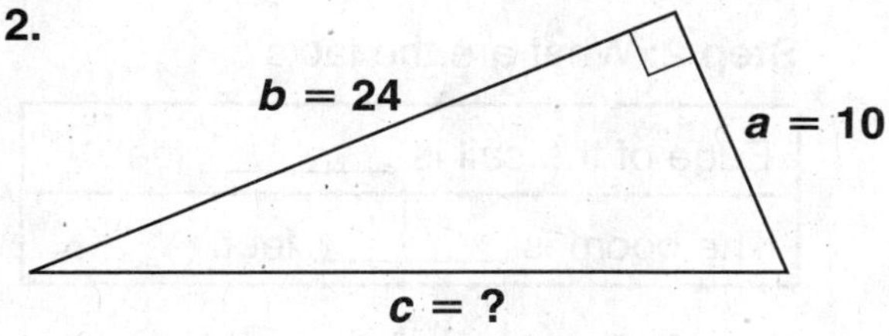

____

3.

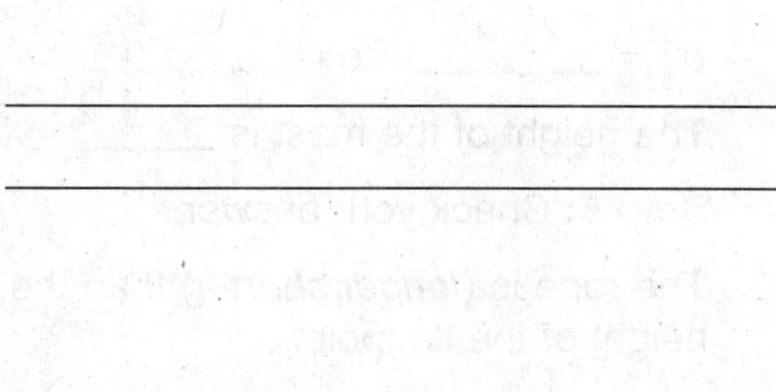

____

____

In questions 4–12, *a* and *b* are the legs of a right triangle and *c* is the hypotenuse. Find the missing side.

4. $a = 4$, $b = 3$, $c =$ ______

5. $a = 15$, $b = 36$, $c =$ ______

6. $a = 20$, $b =$ ______, $c = 25$

7. $a =$ ______, $b = 15$, $c = 17$

8. $a = 16$, $b =$ ______, $c = 34$

9. $a =$ ______, $b = 24$, $c = 26$

10. $a = 6$, $b =$ ______, $c = 10$

11. $a = 9$, $b = 12$, $c =$ ______

12. $a = 15$, $b = 20$, $c =$ ______

Name ______________________________ Date ____________ Class ____________

STANDARD 7MG3.3

Focus on CAHSEE Standards: Intervention

Problem Solving Practice

Grade 7 • Measurement and Geometry 3.3

Know and understand the Pythagorean theorem and its converse and use it to find the length of the missing side of a right triangle and the lengths of other line segments and, in some situations, empirically verify the Pythagorean theorem by direct measurement.

1. A flagpole is raised at camp and is supported by three ropes as shown. What is the length, in feet, of each rope from the top of the flagpole to the bottom of a stake?

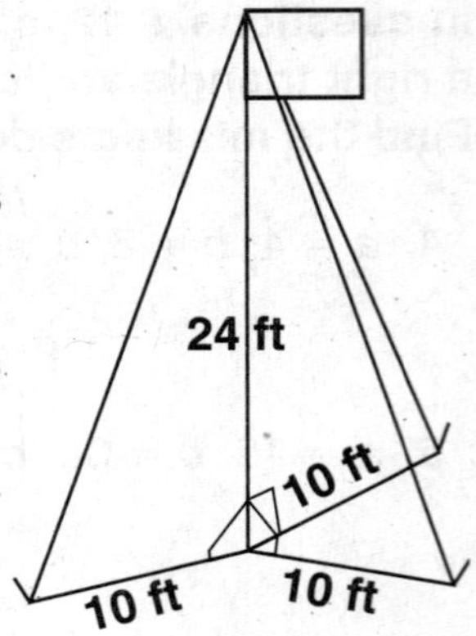

Step 1: What are you asked to find?

____

____

Step 2: What are the facts?

Distance from base of the flagpole to stake is ________ feet.
The flagpole is ________ feet tall.

Step 3: Break down the problem.

Use the Pythagorean theorem.
$a^2 + b^2 = c^2$ $\quad a = 10, b = 24$

Step 4: Solve the problem.

________ $+ 24^2 = c^2$

________ $+$ ________ $= c^2$

________ $= c^2$ ________ $= c$

The length of the rope is ________ feet.

2. The boom and the edge of the sail have lengths as shown. What is the height of the mast?

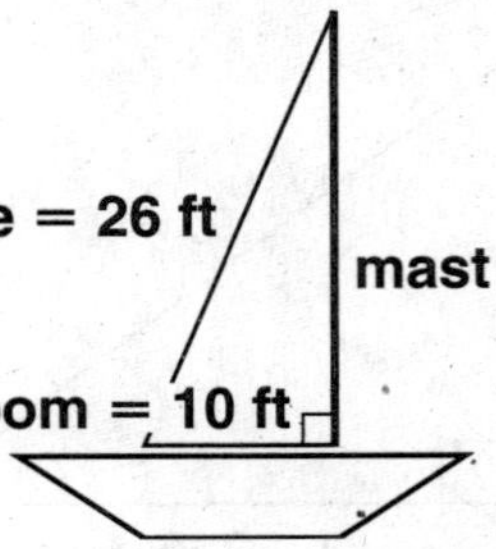

Step 1: What are you asked to find?

____

Step 2: What are the facts?

Edge of the sail is ________ feet.
The boom is ________ feet.

Step 3: Break down the problem.

Use the Pythagorean theorem.
$a^2 + b^2 = c^2$

Step 4: Solve the problem.

________ $+ b^2 =$ ________^{2}

$100 + b^2 = 676$

$b^2 = 676 -$ ________

$b^2 =$ ________ $\quad b =$ ________

The height of the mast is ________ feet.

Step 5: Check your answer.

The rope is (*longer/shorter*) than the height of the flagpole.

The mast is (*longer/shorter*) than the edge of the sail.

Name ______________________ Date ____________ Class ____________

STANDARD 7MG3.4

Focus on CAHSEE Standards: Intervention

Alternative Strategies

Grade 7 • Measurement and Geometry 3.4

Demonstrate an understanding of conditions that indicate two geometrical figures are congruent and what congruence means about the relationships between the sides and angles of the two figures.

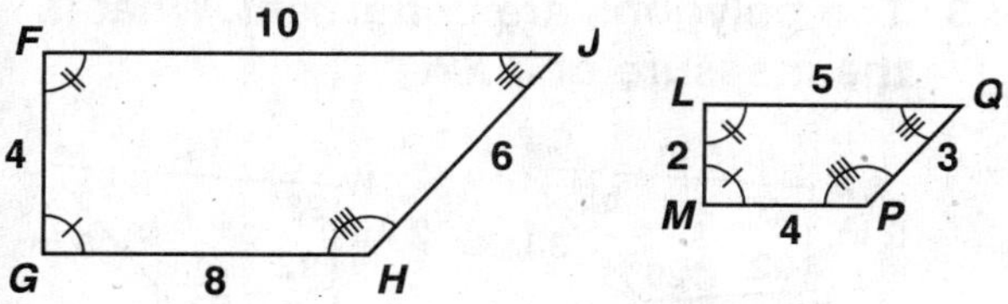

Are the quadrilaterals similar, congruent, or neither?

1a. What are the corresponding angles?

____

____

1b. Are the measures of the corresponding angles equal? ________

2a. List the corresponding sides of the two quadrilaterals.

____

____

2b. What is the ratio of *FG* to *LM*? ______

$\overline{GH}$ to $\overline{MP}$? ______ $\overline{JH}$ to $\overline{QP}$? ______

$\overline{FJ}$ to $\overline{LQ}$? ______

2c. Is there a common ratio for all of the corresponding side lengths? ________

What is the ratio? ______

2d. The quadrilaterals are ____________.

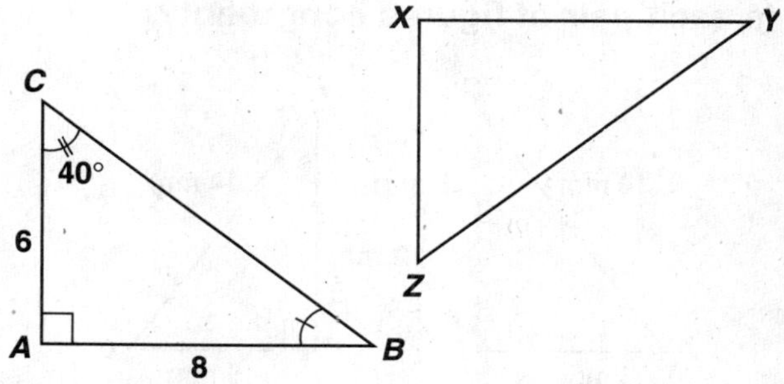

$\triangle ABC$ is congruent to $\triangle XYZ$. Find the measure of $\angle Y$.

3a. The figures are congruent. What can you use to find the dimensions of $\triangle XYZ$?

____

3b. To which angle of $\triangle XYZ$ does each angle of $\triangle ABC$ correspond?

____

____

3c. Since $\angle C$ corresponds to $\angle Z$, what do we know about those angles' measures?

____

4a. What is the sum of the measures of the inside angles of any triangle? ________

$m\angle A +$ ________ $+ m\angle C = 180.$

4b. Substitute the measures of the two known angles in $\triangle ABC$ to calculate $\angle B$.

________ $+ m\angle B +$ ________ $= 180$

$m\angle B = 180 -$ ________ $=$ ________

4c. What is the measure of $\angle Y$? ________

Name ______________________ Date ____________ Class ____________

STANDARD **7MG3.4**

Focus on CAHSEE Standards: Intervention

Sample Problems

Grade 7 • Measurement and Geometry 3.4

Demonstrate an understanding of conditions that indicate two geometrical figures are congruent and what congruence means about the relationships between the sides and angles of the two figures.

Is each pair of figures congruent?

1.

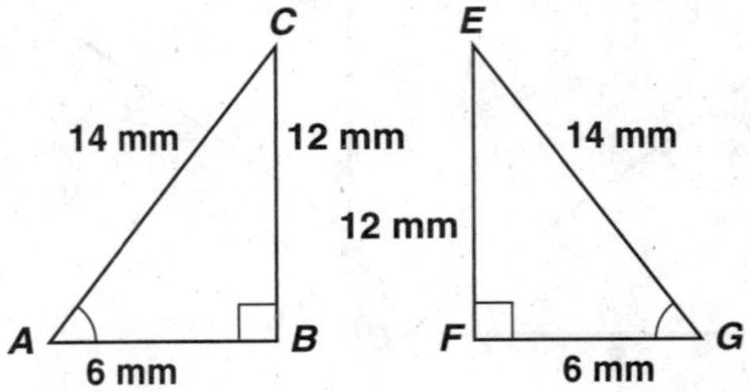

For two figures to be congruent, the measures of the corresponding angles must be equal and the lengths of the corresponding ________ must be equal.

Are the measures of the corresponding angles equal? ________

Are the lengths of corresponding sides equal? ________ Are these two figures congruent? ________

2.

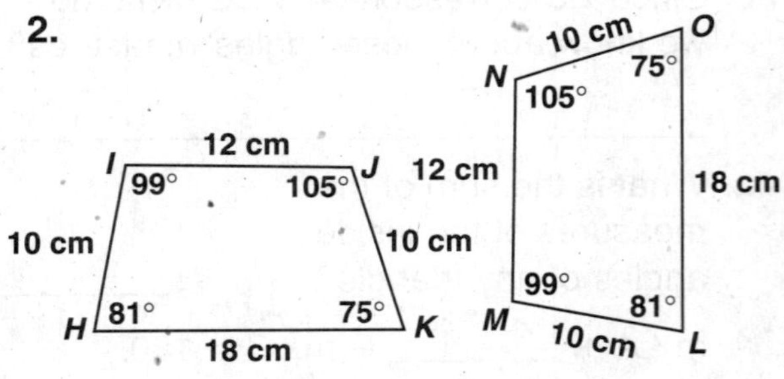

Compare the corresponding parts.

Are the corresponding angles and the lengths of the corresponding sides equal? ________

Are these two figures congruent? ________

3. The polygons are congruent. What is the measure of ∠*X*?

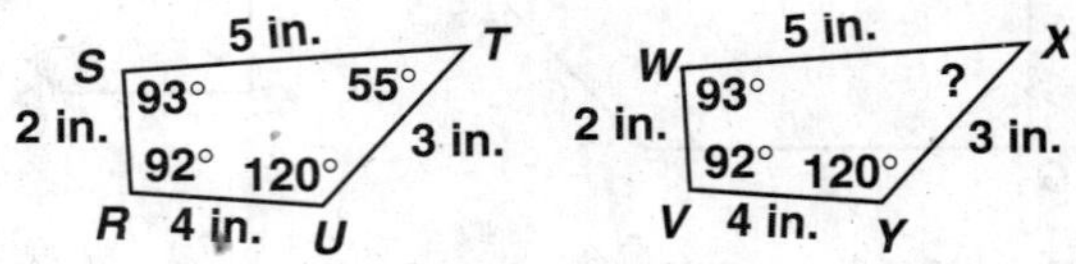

If two figures are congruent, you can use corresponding parts to identify any unlabeled dimensions in the figures.

Which angle in polygon *STUR* corresponds to ∠*X* in polygon *WXYV*? ________

What is the measure of ∠*T*? ________

Because ∠*X* and ∠*T* are corresponding angles of congruent figures,

if ∠*T* = 55°, then ∠*X* = ________.

4. The two pentagons are congruent. What is the missing side length?

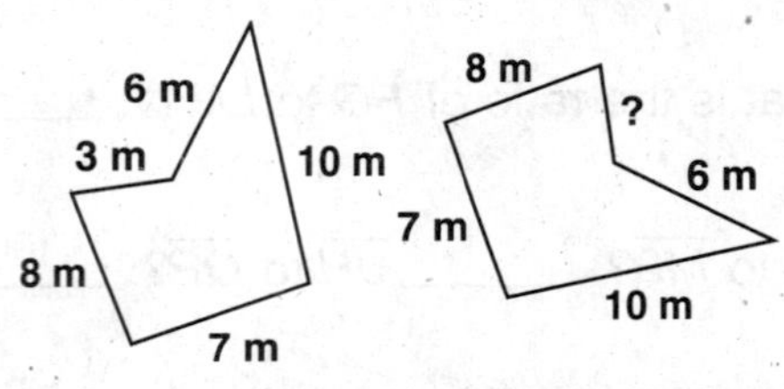

Compare the corresponding sides.

The side for which the measure is missing in the second figure is between sides measuring

________ and ________.

In the first figure, what is the measure of the side between the sides measuring 8 m and 6 m? ________

What is the missing side length in the second figure? ________

Name ______________________ Date ____________ Class ____________

STANDARD 7MG3.4

Focus on CAHSEE Standards: Intervention

Skills Practice

Grade 7 • Measurement and Geometry 3.4

Demonstrate an understanding of conditions that indicate two geometrical figures are congruent and what congruence means about the relationships between the sides and angles of the two figures.

For questions 1–4, tell whether each figure is a polygon? Write *yes* or *no*.

1.

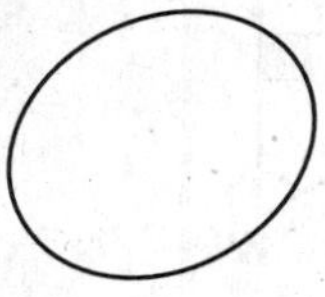

____

2.

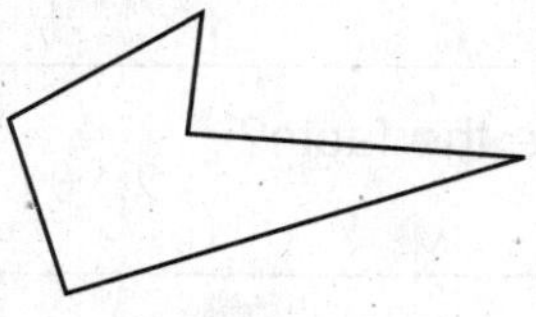

____

3.

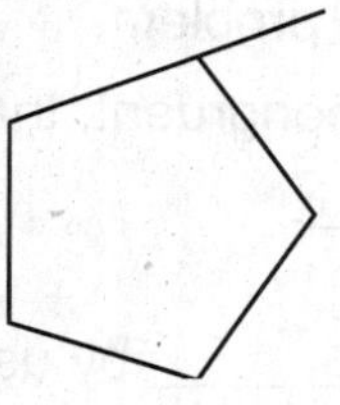

____

4.

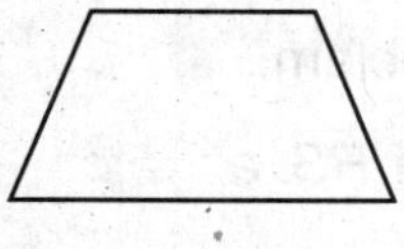

____

For questions 5–7, tell whether each pair of figures is congruent. If not, tell why not.

5.

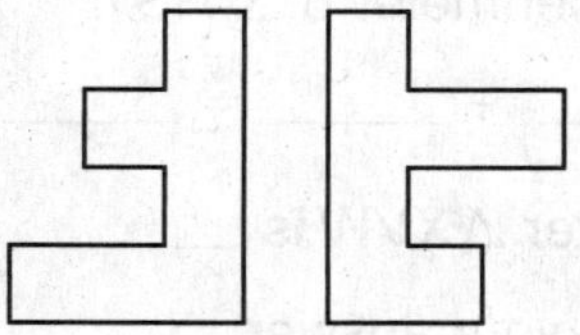

____

6.

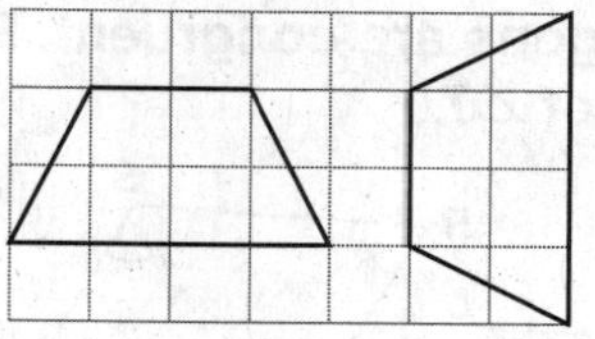

____

7.

____

For questions 8–13, give the measures of the specified angles and sides. The trapezoids are congruent.

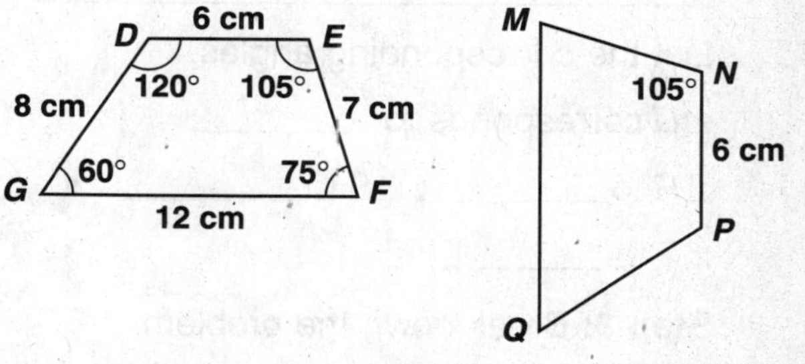

8. m∠*M* = ________.

9. m∠*P* = ________.

10. m∠*Q* = ________.

11. *MN* = ________.

12. *MQ* = ________.

13. *PQ* = ________.

Name ____________________ Date __________ Class __________

Focus on CAHSEE Standards: Intervention

STANDARD 7MG3.4

Problem Solving Practice

Grade 7 • Measurement and Geometry 3.4

Demonstrate an understanding of conditions that indicate two geometrical figures are congruent and what congruence means about the relationships between the sides and angles of the two figures.

1. The two polygons are congruent. Find the measure of ∠*T*.

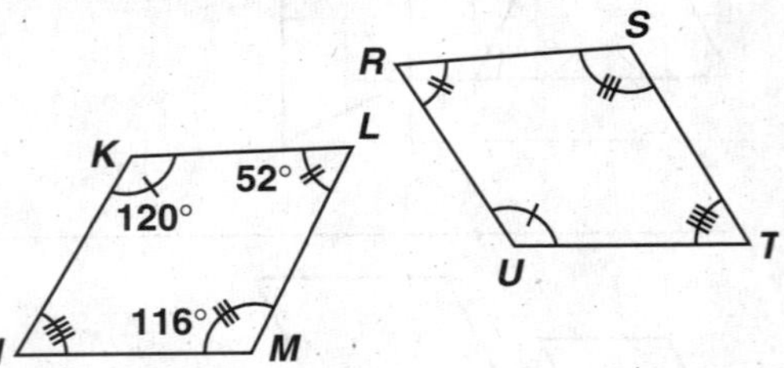

Step 1: What are you asked to find?

____

Step 2: What are the facts?

What is the relationship between the two polygons?

____

List the corresponding angles.

∠*U* corresponds to ________;

∠*R* to ________; ∠*S* to ________;

∠*T* to ________

Step 3: Break down the problem.

Because they are corresponding angles, the m∠*T* = m________.

The sum of the measures of the interior angles of a quadrilateral is ________, so

m∠*K* + m_____ + m_____ + m∠*N* = 360

Step 4: Solve the problem.

_____ + _____ + _____ + m∠*N* = 360

m∠*N* = 360 − ________

m∠*N* = ________

m∠*N* = m∠*T*, so m∠*T* = ________.

Step 5: Check your answer.

2. △*RST* ≅ △*XVW*. What is the perimeter of △*XVW*?

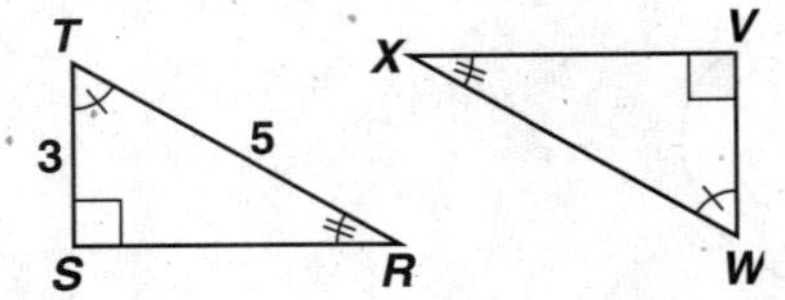

Step 1: What are you asked to find?

____

Step 2: What are the facts?

____

____

Step 3: Break down the problem.

Since the triangles are congruent, the perimeters are ________.

First find the length of ________ by using the ________________ theorem.

Step 4: Solve the problem.

Calculate the length of $\overline{RS}$: $a^2 = c^2 - b^2$

$a^2 =$ _____ − _____

$=$ _____ − _____ = _____

Length of $\overline{RS}$ = _____

Calculate the perimeter of △*RST*.

_____ + _____ + _____ = _____

So the perimeter △*XVW* is ______.

Step 5: Check your answer.

Name ______________________ Date __________ Class __________

Focus on CAHSEE Standards: Intervention

STANDARD 7PS1.1 *Sample Problems*

Grade 7 • Statistics, Data Analysis, and Probability 1.1

Know various forms of display for data sets, ~~including a stem-and-leaf plot or box-and-whisker plot~~; use the forms to display a single set of data or to compare two sets of data.

Use the line graph to answer the question.

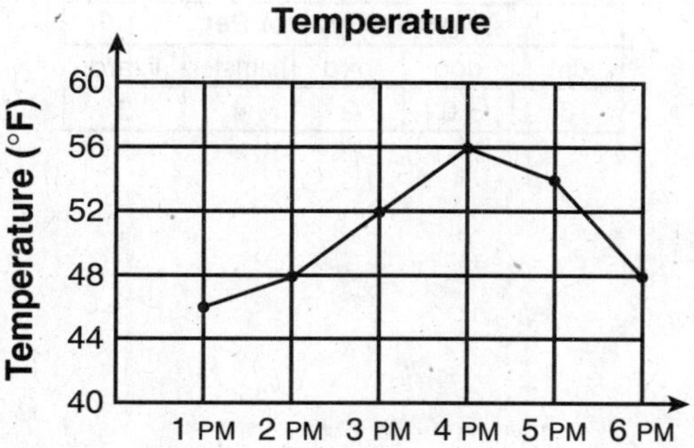

1. Which is greater, the change in temperature between 4:00 PM and 5:00 PM or between 5:00 PM and 6:00 PM?

What is the temperature at 4:00 PM?

________ At 5:00 PM? ________

Find the difference:

56 − ________ = ________

What is the temperature at 6:00 PM?

____

Find the difference between the temperatures at 5:00 PM and 6:00 PM:

________ − 48 = ________

Compare the changes.

2°F ________ 6°F

Between which hours is the change greater?

____

Use the double bar graph to answer the question.

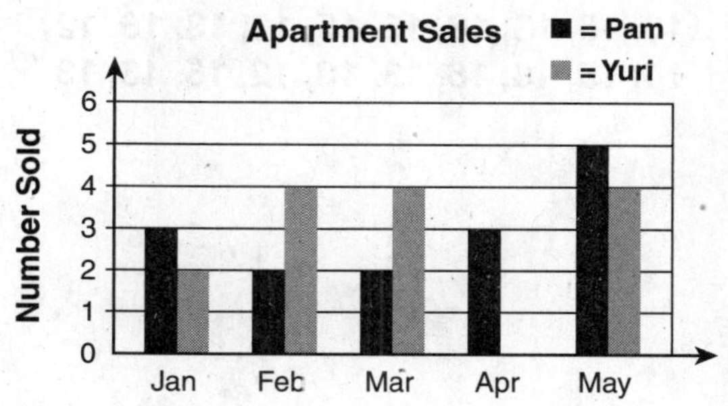

2. Who sold more apartments between January and May, Pam or Yuri?

How is the data for Pam's apartment sales shown? ________________

The data for Yuri's sales?

____

How many apartments did Pam sell in each of the 5 months?

____

____

Find Pam's total:

3 + 2 + ____ + ____ + ____ = ______

How many apartments did Yuri sell in each of the 5 months?

____

____

Find Yuri's total:

2 + 4 + ____ + ____ + ____ = ______

Who sold more apartments between January and May? ________

Name ______________________ Date __________ Class __________

STANDARD 7PS1.1

Focus on CAHSEE Standards: Intervention

Skills Practice

Grade 7 • Statistics, Data Analysis, and Probability 1.1

Know various forms of display for data sets, ~~including a stem-and-leaf plot or box-and-whisker plot~~; use the forms to display a single set of data or to compare two sets of data.

1. **Make a line plot of the following student ages:**

   16, 15, 13, 16, 16, 15, 12, 13, 16, 12, 14, 15, 16, 16, 13, 10, 12, 15, 13, 13

   How many students were older than 13?

   ____

2. **Make a line graph of the snowfall data shown in the table.**

Monthly Snowfall (inches)					
Oct	Nov	Dec	Jan	Feb	Mar
0	4	16	10	24	6

   Between which two months was the difference in the amount of snow the greatest?

   ____

3. **Make a circle graph of the survey data shown in the table.**

Favorite Type of Pet				
cat	dog	bird	hamster	lizard
6	10	2	4	3

   Which pet was the favorite pet for about $\frac{1}{4}$ of the people surveyed? __________

4. **Make a double bar graph of the data shown in the table.**

	Posters Made				
	Mon	Tue	Wed	Thu	Fri
Tam	3	6	6	8	10
Jon	4	7	6	2	10

   On which day did Tam make more than double the number of posters John made?

   ____

Name ______________________ Date ___________ Class ___________

Focus on CAHSEE Standards: Intervention

STANDARD 7PS1.2

Sample Problems

Grade 7 • Statistics, Data Analysis, and Probability 1.2

Represent two numerical variables on a scatter plot and informally describe how the data points are distributed and any apparent relationship that exists between the two variables (e.g., between time spent on homework and grade level).

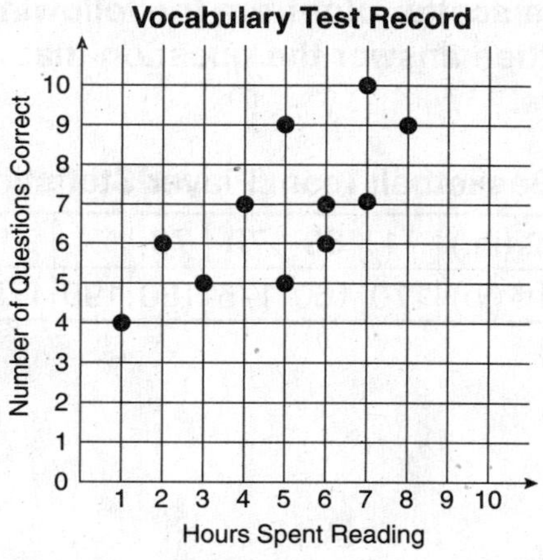

Jill created this scatter plot to compare the number of hours she spent reading to the number of questions she answered correctly each week on a 10-question vocabulary test.

1. How many questions did she get right when she spent 4 hours reading?

Find the line for 4 hours of reading.

Follow that line until it intersects with a set value. Then follow the intersecting line to find out how many questions she answered correctly. ______

2. Jill got 9 questions right twice. How many hours did she read before each of those tests?

Find the line for 9 questions correct.

Follow that line until it intersects with a set value. Then follow the intersecting line to find out how many hours she read. ______

3. In general, as Jill read more, how did it affect her test scores?

____

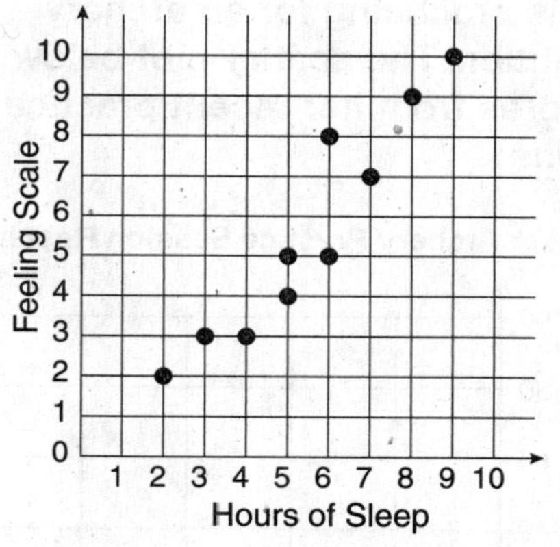

Paulo created this scatter plot to show how he felt after sleeping for a certain number of hours. He used a feeling scale of 1 to 10, with 1 meaning tired and 10 meaning alert.

4. Do the two data sets have a positive correlation, a negative correlation, or no correlation?

In what direction does the line of best fit go? ______

Does the Feeling Scale go up with the more sleep Paulo gets? ______

What is the correlation? __________

When did Paulo feel the most alert?

____

Name _______________________________ Date ____________ Class ____________

STANDARD 7PS1.2

Focus on CAHSEE Standards: Intervention

Skills Practice

Grade 7 • Statistics, Data Analysis, and Probability 1.2

Represent two numerical variables on a scatter plot and informally describe how the data points are distributed and any apparent relationship that exists between the two variables (e.g., between time spent on homework and grade level).

Avery is practicing for an archery competition. The scatter plot below shows her scores from her recent practice sessions.

Archery Practice Session Results

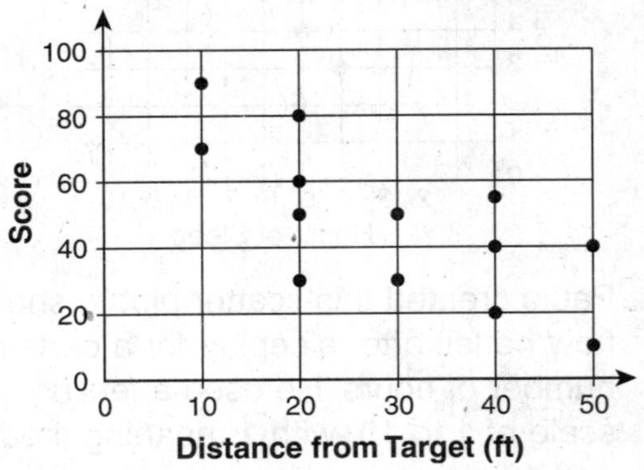

1. What was Avery's highest score when she shot at a target 30 feet away? ______

2. What was the second-highest score Avery got? How far was she from her target at the time? ______________

3. What were Avery's scores when she shot at a target 40 feet away?

____

4. In general, what happened to Avery's scores as she got farther away from the target?

____

5. How would you describe the correlation between the two sets of data?

____

Make a scatter plot from the following data, then answer the question that follows.

Basketball Team Player Statistics

Height (in.)	71	68	70	73	74	71	74
Weight (lb)	170	160	175	180	190	175	170

6. How many pounds did the tallest players weigh? ______________

7. Does there seem to be a correlation between height and weight? ______

8. How would you describe the correlation between the two data sets?

____

Name ______________________ Date __________ Class __________

STANDARD 1A2.0

Focus on CAHSEE Standards: Intervention

Alternative Strategies

Algebra 1 • Standard 2.0

Students understand and use such operations as taking the opposite, finding the reciprocal, and taking a root ~~and raising to a fractional power~~. They understand and use the rules of exponents.

Rewrite each product in exponential form.

1. 4 • 4

What is the factor that is multiplied by itself?

___

How many times does the factor appear? ___

How can this product be written as a base and an exponent? ___

2. 5 • 5 • 5 • 5 • 5

What is the factor that is multiplied by itself?

___

How many times does the factor appear? ___

How can this product be written as a base and an exponent? ___

Solve for the unknown.

3. $x^3 = 8$

How could you rewrite this equation with the exponential term as a product?

____

Which value satisfies this equation? $x =$ ___

4. $t^2 = 81$

How could you rewrite this equation with the exponential term as a product? __________

Which positive value satisfies this equation?

$t =$ ______

Simplify.

5. $\sqrt{5}$

What does the radical symbol, "$\sqrt{\ }$", indicate?

____

____

Is 9 a perfect square? ______

What positive number times itself is equal to 9? ___

6. $\sqrt{50}$

Is 50 a perfect square? ______

Does 50 have a perfect square as a factor?

____

Factor the original radical to include that perfect square.

____

How can you express this as the product of two radicals?

____

What is the final simplified version of the radical? ______

Name ______________________ Date ____________ Class ____________

STANDARD 1A2.0

Focus on CAHSEE Standards: Intervention

Sample Problems

Algebra 1 • Standard 2.0

Students understand and use such operations as taking the opposite, finding the reciprocal, and taking a root ~~and raising to a fractional power~~. They understand and use the rules of exponents.

Simplify using the rules of exponents.

1. $5^3 \cdot 5^4$

When multiplying two exponential terms with the same base, you should ______ the exponents.

Write the equivalent of $5^3 \cdot 5^4$ using a single base and exponent: ___

2. $(3^2)^3 =$

When you raise an exponential term to a power, you should _________ the exponents.

Write the equivalent of $(3^2)^3$ using a single base and exponent: ___

3. $(y^3 \cdot -3xz)^2$

What rule of exponents applies here?

____

____

$(y^3 \cdot -3xz)^2 =$ _________

4. $4^0 =$

Any nonzero number to the zero power is equal to what? ___

5. $\sqrt{3}(5 - \sqrt{2}) =$

Which property can you use to simplify this expression? ______________

Write the final simplified result: ___________

Simplify.

6. $\sqrt{\frac{25}{8}} =$

What must your first step be?

____

____

____

You can eliminate the radical in the denominator by multiplying by 1 in what form?

______

Write the final simplified result: ______

7. 4^{-3}

Write this expression with a positive exponent. _________

Write the final simplified result: _______

Solve.

8. $2\sqrt{x - 4} = 6$

To solve for x, what operation must you first perform to isolate the radical?

____

How can you then eliminate the radical?

____

What is the final step to isolate the variable?

____

$x =$ ___

Name ______________________ Date __________ Class __________

STANDARD 1A2.0

Focus on CAHSEE Standards: Intervention

Skills Practice

Algebra 1 • Standard 2.0

Students understand and use such operations as taking the opposite, finding the reciprocal, and taking a root ~~and raising to a fractional power~~. They understand and use the rules of exponents.

Simplify using the rules of exponents.

1. $(2^3)^5 =$ ______

2. $8^9 \cdot 8^3 =$ ______

3. $(3 \cdot 6)^0 =$ ______

4. $(abc^2)^3 \cdot ab =$ ______

5. $5^{-3} =$ ______

Simplify.

6. $\sqrt{56} =$ ______

7. $\sqrt{\frac{4}{25}} =$ ______

8. $9\sqrt{100} =$ ______

9. $-4\sqrt{\frac{1}{10}} =$ ______

10. $\frac{3x^{-3}}{x} =$ ______

Name ________________ Date ________ Class ________

Focus on CAHSEE Standards: Intervention

STANDARD 1A2.0

Skills Practice continued

Algebra 1 • Standard 2.0

Students understand and use such operations as taking the opposite, finding the reciprocal, and taking a root ~~and raising to a fractional power~~. They understand and use the rules of exponents.

Simplify.

11. $4 + \sqrt{3} + 2\sqrt{3} =$

____

12. $3\sqrt{6} + \sqrt{24} =$

____

13. $\sqrt{3}(5\sqrt{3} - 2\sqrt{6}) =$

____

14. $\frac{6}{\sqrt{10}} =$

____

15. $\sqrt{64} + \sqrt{4} =$

____

16. $6^{-2} =$

____

17. $\sqrt{9} \div 3^{-1} =$

____

Solve.

18. $14 = \sqrt{x}$

$x =$ ________

19. $\sqrt{5x + 1} + 8 = 12$

$x =$ ________

20. $8\sqrt{x + 3} = 64$

$x =$ ________

Name ______________________ Date ____________ Class ____________

STANDARD **1A3.0**

Focus on CAHSEE Standards: Intervention

Sample Problems

Algebra 1 • Standard 3.0

Students solve equations and inequalities involving absolute values.

1. What is the absolute value of a number? Give an example.

____

____

____

Example: $|10| = 10$ ______________

2. Solve $|x + 15| = 3$.

The given equation is an absolute value equation.

What is its general form? ____________

Can it be solved?

____

____

____

What are these two equations?

____

Solve both equations

$x + 15 = 3$ and $x + 15 = -3$

$x + 15 - 15 = 3 - 15$ and $x + 15 - 15 = -3 - 15$

$x =$ _____ and $x =$ _____

The solutions are $x =$ _____, and $x =$ _____.

3. Solve $|x - 21| = -8$.

What is the value of c? _____

What happens when the c value is negative? ____________

____

4. Solve $|3x + 3| < 6$.

What is the general form of this inequality? ____________

____

____

What are the values of a, b and c?

____

What are the two equations?

____

Solve $3x + 3 < 6$ and $3x + 3 > -6$.

The solution is ____________

Name ____________________ Date __________ Class __________

STANDARD 1A3.0

Focus on CAHSEE Standards: Intervention

Skills Practice

Algebra 1 • Standard 3.0

Students solve equations and inequalities involving absolute values.

1. Solve $|w| > 7$.

2. Solve $|x - 3| = 20$.

3. Solve $|x + 21| = -9$.

4. Solve $|2 - w| + 3 = 5$.

5. Solve $|y + 4| = -8$.

6. Solve $7 + |t + 3| = 10$.

7. Solve $|z + 1| < 2$.

8. Solve $|x - 5| > 2$.

9. Solve $|x - 7| < -2$.

10. Solve $|x - 3| \leq 1$.

11. Solve $|x - 4| \leq 4$.

12. Solve $|2x - 15| \leq -5$.

13. Solve $|4 + 3x| \leq 13$.

14. Solve $|x - 6| - 5 \leq 10$.

Name ______________________________ Date ____________ Class ____________

STANDARD **1A4.0**

Focus on CAHSEE Standards: Intervention

Alternative Strategies

Algebra 1 • Standard 4.0

Students simplify expressions before solving linear equations and inequalities in one variable, such as $3(2x - 5) + 4(x - 2) = 12$.

Identify and apply the appropriate properties of numbers.

1. $a + 2 + 9a$

The ______________ Property allows you to rewrite the order of the terms that are being added or multiplied without changing the result.

Rewrite the expression above so that like terms are next to each other.

____

2. $5 \cdot (4 \cdot b)$

The ______________ Property allows you to regroup numbers in addition or multiplication to change the order in which something is calculated.

Rewrite the expression above so that like terms are grouped together by parentheses.

____

3. $4(y + 1)$

The ______________ Property of Multiplication allows you to distribute a factor *outside* the parentheses to the terms *inside* the parentheses.

Eliminate the parentheses and rewrite the expression above as the sum of two products.

____

Use the appropriate properties of numbers to simplify.

4. $13a - 8 + 3(6a - 2)$

First apply the ______________ Property to distribute the factor through the parentheses.

$13a - 8 + 3($______$) + 3($______$)$

$13a - 8 +$ ______ $-$ ______

You can rewrite the order of these terms

because of the ______________ Property.

____

Simplify by combining like terms.

____

5. $-(3x - 2)$

What factor is outside the parentheses?

______

Distribute the factor outside the parentheses. Then simplify the expression.

____

Name ______________________ Date __________ Class __________

STANDARD 1A4.0

Focus on CAHSEE Standards: Intervention

Sample Problems

Algebra 1 • Standard 4.0

Students simplify expressions before solving linear equations and inequalities in one variable, such as $3(2x - 5) + 4(x - 2) = 12$.

Simplify each equation or inequality. Do not solve.

1. $5(10x - 6) = 4$

What property of numbers will allow you to distribute the 5 through the expression inside the parentheses?

____

____

Distribute the 5.

Result: ______________________

2. $x(9 - 6) + 9(x - 2) > 80$

What operation should you complete first?

____

____

Result: ______________________

Distribute the 9.

Result: ______________________

Combine like terms.

Result: ______________________

3. $9(7a - 4) = 5a - 2(a + 1)$

Apply the Distributive Property of Multiplication.

Result: ______________________

Combine like terms.

Result: ______________________

4. $3c + 5(4c - 2) > -(c - 6) + 2c$

Apply the Distributive Property of Multiplication.

Result: ______________________

____

Combine like terms.

Result: ______________________

Name ______________________ Date __________ Class __________

STANDARD 1A4.0

Focus on CAHSEE Standards: Intervention

Skills Practice

Algebra 1 • Standard 4.0

Students simplify expressions before solving linear equations and inequalities in one variable, such as $3(2x - 5) + 4(x - 2) = 12$.

For questions 1–3, identify like terms in each expression.

1. $z + 3 - 3z$

____

2. $2p + 8p + 7$

____

3. $-6w + 4w - 3$

____

For questions 4–12, simplify each expression.

4. $12(x - 3)$

____

5. $-(a - 9)$

____

6. $-3(5a + 9)$

____

7. $12(3x - 2)$

____

8. $x(23 + 2)$

____

9. $-3(y - 7)$

____

10. $2x - 8x - 12 + 8x$

____

11. $7(x + 3) - 2(5 - x)$

____

12. $2x + 3x + 2 - 8x$

____

Name ______________________ Date ____________ Class ____________

STANDARD 1A4.0

Focus on CAHSEE Standards: Intervention

Skills Practice continued

Algebra 1 • Standard 4.0

Students simplify expressions before solving linear equations and inequalities in one variable, such as $3(2x - 5) + 4(x - 2) = 12$.

Simplify each expression.

13. $2x + 3 + 23x$

____

14. $2(23a + 12)$

____

15. $a + 10 + a - 6$

____

16. $12 - 3m - 12m + 3m$

____

17. $5(x + 3) - 2(4x + 1)$

____

18. $-4(2k - 2) - 3k$

____

19. $2x(3) + 3(x + 2)$

____

20. $4(x + 2) - (x + 2)$

____

21. $9 - 4(9 + y)$

____

22. $-5(2 + 7x) - 3x$

____

23. $y - 2y + 3(y + 1)$

____

24. $11(2j + 4) - 11j$

____

Name ______________________ Date ____________ Class ____________

Focus on CAHSEE Standards: Intervention

STANDARD 1A5.0

Alternative Strategies

Algebra 1 • Standard 5.0

Students solve multistep problems, including word problems, involving linear equations and linear inequalities in one variable and provide justification for each step.

Use algebra tiles to solve $x - 5 = 9$.

Let x = [+]

Let [–] = −1 and [+] = 1

Show the equation.

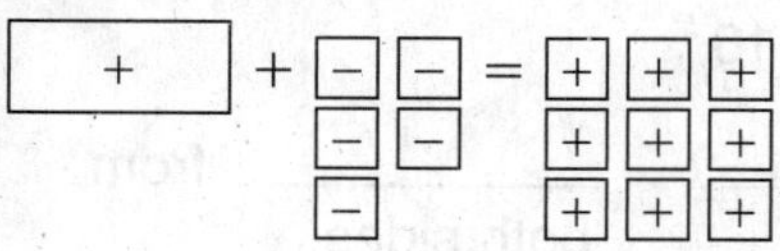

Remember [+] [–] = 0 or 1 negative and 1 positive tile cancel each other out.

Add [+] [+] [+] [+] [+] to both sides of the equation.

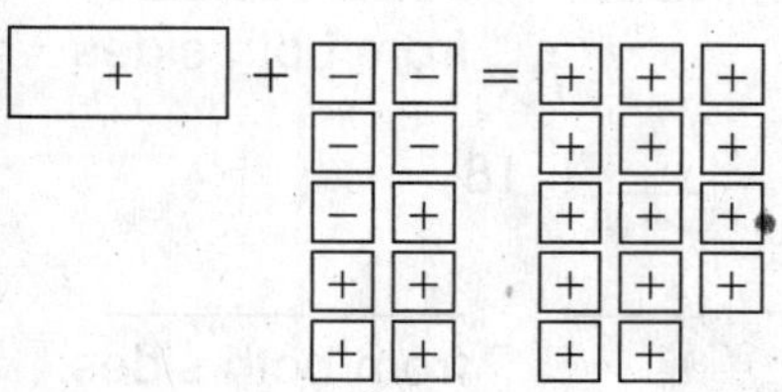

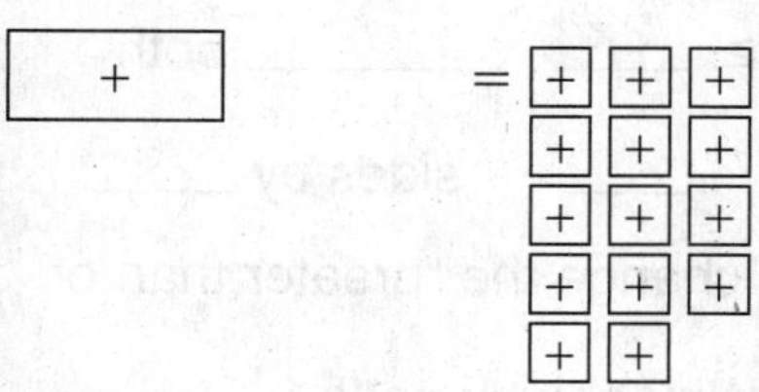

$x = 14$

Use algebra tiles to solve $2x - 3 = 5$.

Let x = [+]

Let [–] = −1 and [+] = 1

Show the equation.

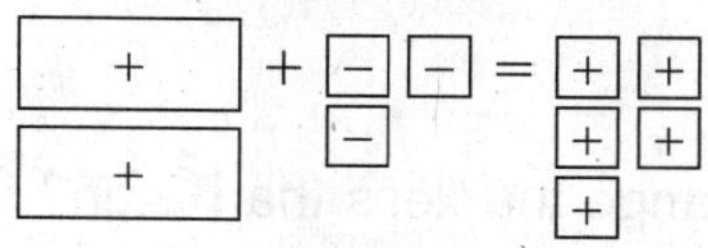

Remember [+] [–] = 0 or 1 negative and 1 positive tile cancel each other out.

Add [+] [+] [+] to both sides of the equation.

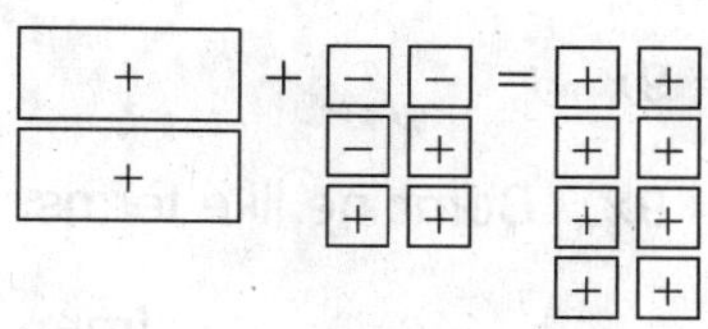

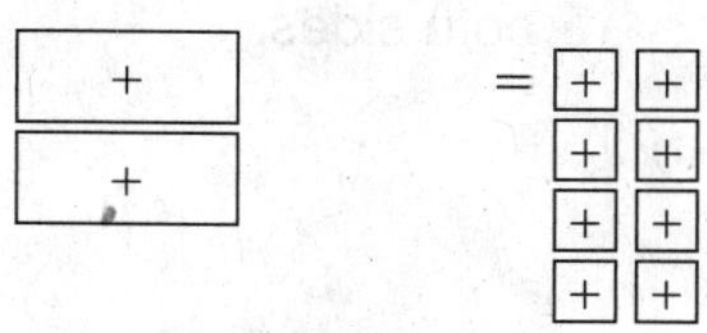

$x = 4$

Name ______________________ Date __________ Class __________

Focus on CAHSEE Standards: Intervention

STANDARD 1A5.0

Sample Problems

Algebra 1 • Standard 5.0

Students solve multistep problems, including word problems, involving linear equations and linear inequalities in one variable and provide justification for each step.

Solve.

1. $x - 5 = 10$

_____ _____ _______ to both sides.

$x =$ _____

2. $-4x < 1$

$\frac{-4x < 1}{\_\_\_\_ \quad \_\_\_\_}$ _______ both sides by _____.

Do you need to change the "less than" sign?

Why or why not? ______________________

____

____

____

x __________

3. $2x + 6x - 4 = 9x$

_____ $- 4 = 9x$ Combine like terms.

_____ _____ __________ from both sides.

_____ $= x$

4. $4x + 8 > 24$

_____ _____ __________ from both sides.

$\frac{4x}{\_\_\_\_} > \frac{\_\_\_\_}{\_\_\_\_}$ _______ both sides by _____.

Do you need to change the "greater than" sign? Why or why not? ______________________

____

____

x __________

5. $2x + 12 = -4x + 18$

_____ _____ _______ to both sides.

$6x + 12 = 18$

_____ _____ __________ from both sides.

$\frac{6x}{\_\_\_\_} = \frac{6}{\_\_\_\_}$ _______ both sides by _____.

$x =$ __________

6. $3(x + 2) \geq 6x + 18$

___ + ___ $\geq 6x + 18$ __________ the 3.

_____ _____ __________ from both sides.

$6 \geq$ ___ $+ 18$

_____ _____ __________ from both sides.

$\frac{\_\_\_\_}{\_\_\_\_} \geq \frac{3x}{\_\_\_\_}$ _______ both sides by _____.

Do you need to change the "greater than or equal to" sign? Why or why not? __________

____

____

__________ x

Name ______________________ Date ____________ Class ____________

STANDARD 1A5.0

Focus on CAHSEE Standards: Intervention

Skills Practice

Algebra 1 • Standard 5.0

Students solve multistep problems, including word problems, involving linear equations and linear inequalities in one variable and provide justification for each step.

Solve.

1. $d + 4 = 12$

____

2. $-1 = c - (-12)$

____

3. $4 + 2x = 66$

____

4. $b - 3 = -12$

____

5. $3x = 24$

____

6. $4 = \frac{2}{3}z$

____

7. $288 = 16m$

____

8. $\frac{1}{3}(x - 1) = 6$

____

9. $8a - 3a - 4 = 21$

____

10. $\frac{2}{3}(b + 3) = 6$

____

11. $24 - 6x = 6(4 + x)$

____

12. $6g - 5 = 7g + 7$

____

13. $3(z + 7) = 123$

____

14. $5y - 3y + 4 = 3y + 8$

____

Name ______________________________ Date ____________ Class ____________

Focus on CAHSEE Standards: Intervention

STANDARD 1A5.0

Problem Solving Practice

Algebra 1 • Standard 5.0

Students solve multistep problems, including word problems, involving linear equations and linear inequalities in one variable and provide justification for each step.

1. Sara is building a pen for her pet rabbit. She has 180 feet of fencing. She wants the pen to be twice as long as it is wide. What are the dimensions of the pen?

Step 1: What are you asked to find? ________

____

Step 2: What are the facts?

Sara has ________ feet of fencing. The pen should be ________ as long as it is wide.

Step 3: Break down the problem. Draw a diagram of the pen.

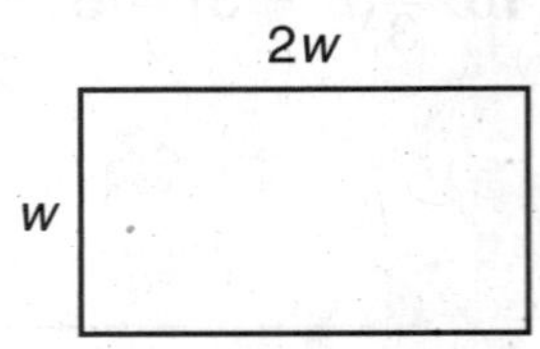

If the pen's width is w, its length = _____.

Write an equation for the pen's perimeter.

$2(\_\_\_\_\_) + 2(\_\_\_\_\_) = P$

Step 4: Solve the problem.
Substitute the known value into the equation and solve for w.

$2w + 2(2w) =$ _____

$2w +$ _____ $= 180$

_____ $= 180$

$w =$ _____

The pen is 30 ft wide and _____ ft long.

Step 5: Check that your answer makes sense.

$2(\_\_\_\_\_) + 2(\_\_\_\_\_) = 180$

2. The glee club sold 200 concert tickets and collected \$640. A student ticket cost \$2 and an adult ticket cost \$5. How many of each type of ticket were sold?

Step 1: What are you asked to find? ________

____

____

Step 2: What are the facts?

Student tickets cost ____. Adult tickets cost ____.

____ tickets were sold in all for a total ____.

Step 3: Break down the problem.
If x = number of adult tickets sold, then $200 - x =$ ________________________.

The total value of the adult tickets is 5______.

The total value of the student tickets is 2 __________.

Write an equation to represent the situation.

$5x + 2(200 - x) =$ _____

Step 4: Solve the problem.

$5x +$ _____ $- 2x = 640$

_____ $+ 400 = 640$

$3x =$ _____

$x =$ _____

There were _____ adult tickets and _____ student tickets sold.

Step 5: Check that your answer makes sense.

$5(\_\_\_\_\_) + 2(\_\_\_\_\_) = 640$

Name ______________________________ Date ____________ Class ____________

STANDARD **1A6.0**

Focus on CAHSEE Standards: Intervention
Alternative Strategies

Algebra 1 • Standard 6.0

Students graph a linear equation and compute the *x*- and *y*- intercepts (e.g., graph $2x + 6y = 4$). ~~They are also able to sketch the region defined by linear inequality (e.g., they sketch the region defined by $2x + 6y < 4$).~~

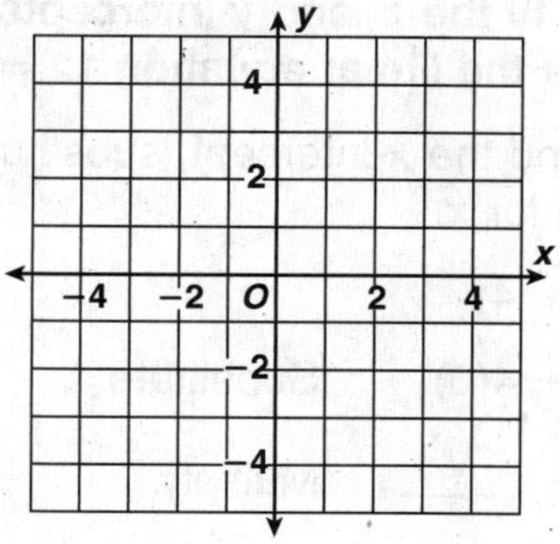

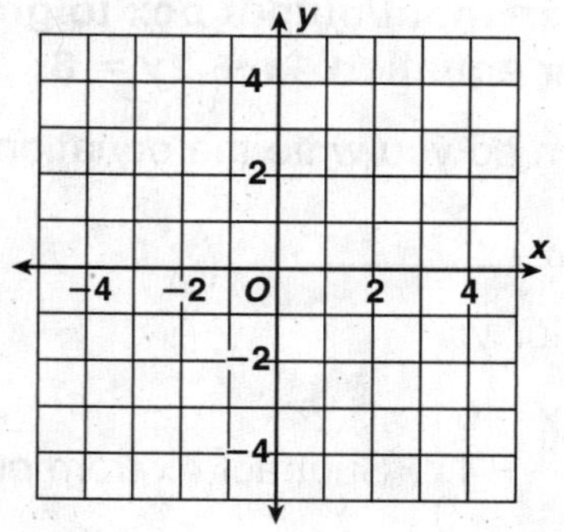

Use the table and the grid to solve.

1. Graph the line for $y = x + 3$.

a. Complete the table.

x	0	1	2	−1	____	____
y	3	____	____	____	1	0

b. Graph the points for the coordinates in the table.

c. The *x*-intercept is the point where the line crosses the ________.

What is the *x*-intercept? ________

d. The *y*-intercept is the point where the line crosses the ________.

What is the *y*-intercept? ________

Use the *x*- and *y*-intercepts and the grid to solve.

2. Graph the line for $y = -3x - 2$.

a. To identify the *x*-intercept, substitute 0 for *y* and solve for *x*.

____ $= -3x - 2$ Substitute 0 for *y*.

____ $= -3x$ Add 2 to both sides.

$\frac{2}{\_\_\_\_} = x$ Divide both sides by $-3x$.

What is the *x*-intercept? ________

b. To identify the *y*-intercept, substitute 0 for *x* and solve for *y*.

$y = -3($____$) - 2$ Substitute 0 for *x*.

$y =$ ____ $- 2$ Multiply.

$y =$ ____ Subtract.

What is the *y*-intercept? ________

c. Plot the intercepts on the grid and graph the line.

Name ______________________________ Date ____________ Class ____________

Focus on CAHSEE Standards: Intervention

STANDARD **1A6.0**

Sample Problems

Algebra 1 • Standard 6.0

Students graph a linear equation and compute the *x*- and *y*- intercepts (e.g., graph $2x + 6y = 4$). ~~They are also able to sketch the region defined by linear inequality (e.g., they sketch the region defined by $2x + 6y < 4$)~~.

1. Use an input/output box to graph the linear equation $4x + 2y = 6$.

What form do you write the equation in?

____

To solve for *y*:

$4x + 2y = 6$

$-4x \qquad -4x$ Subtract 4*x* from both sides.

$2y =$ ______________

$\frac{2y}{2} = \frac{6}{2} - \frac{4x}{2}$ Divide both sides by 2.

$y = 3 - 2x$

Simplify and rearrange in slope intercept form.

$y =$ ______________

Substitute the values of *x* into $y = -2x + 3$.

$y = -2($____$) + 3 =$ ____

$y = -2($____$) + 3 =$ ____

$y = -2($____$) + 3 =$ ____

x	−1	0	1
y	____	____	____

Graph the points for the values in the input/output table.

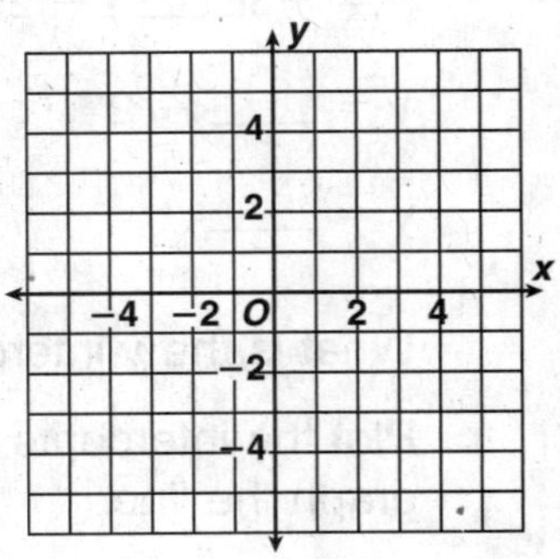

2. Identify the *x*- and *y*-intercepts to graph the linear equation $12 = 3x + 4y$.

First, to find the *x*-intercept, substitute 0 for *y* and solve for *x*.

$12 = 3x + 4y$

$12 = 3x + 4(0)$ Substitute.

$12 = 3x +$ ____ Multiply.

$12 =$ ____ Add.

$\frac{12}{3} = \frac{3x}{3}$ Divide by ____.

____ $= x$ Simplify.

The *x*-intercept is ________.

Now, to find the *y*-intercept, substitute 0 for *x* and solve for *y*.

$12 = 3x + 4y$

$12 = 3(0) + 4y$ Substitute.

$12 =$ ____ $+ 4y$ Multiply.

$12 =$ ____ Add.

$\frac{12}{4} = \frac{4y}{4}$ Divide by ____.

____ $= y$ Simplify.

The *y*-intercept is

________.

Plot the *x*- and *y*-intercepts on the grid. Graph the line they define.

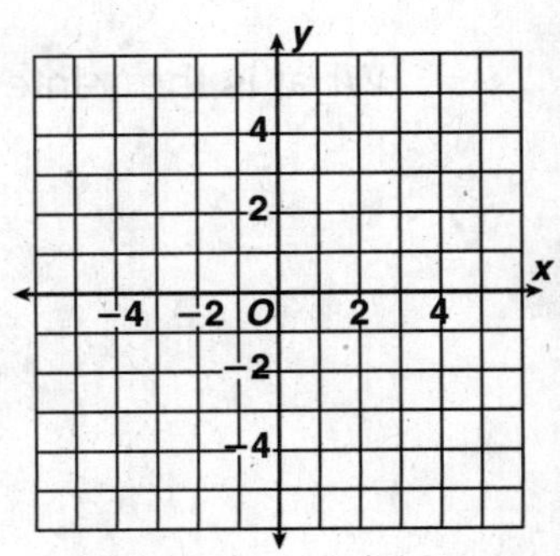

Name ______________________ Date ____________ Class ____________

STANDARD 1A6.0

Focus on CAHSEE Standards: Intervention

Skills Practice

Algebra 1 • Standard 6.0

Students graph a linear equation and compute the x- and y- intercepts (e.g., graph $2x + 6y = 4$). ~~They are also able to sketch the region defined by linear inequality (e.g., they sketch the region defined by $2x + 6y < 4$).~~

Complete an input/output table for the following equations.

1. $y = x + 5$

x	0	1	2	3	4	5
y	____	____	____	____	____	____

2. $2y + 4x = 12$

x	0	1	2	3	4	5
y	____	____	____	____	____	____

Find the *x*- and *y*-intercepts for the graphs of the following equations.

3. $y = 2x + 4$

x-intercept = ____; y-intercept = ____

4. $18 = 2x + 6y$

x-intercept = ____; y-intercept = ____

Use an input/output table to graph the following equations.

5. $15 = 5x - 3y$

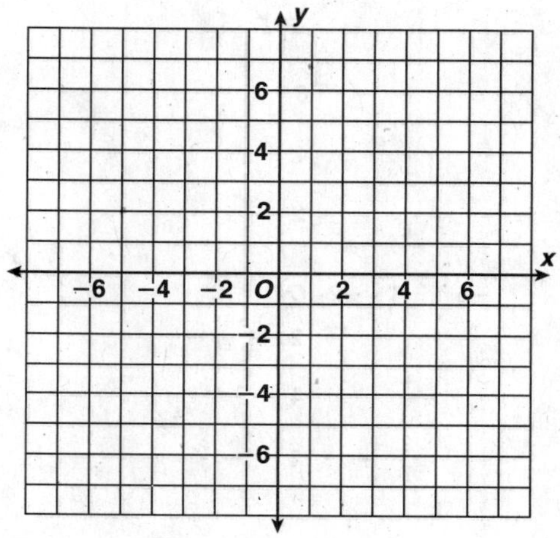

6. $-2x - 4y = 8$

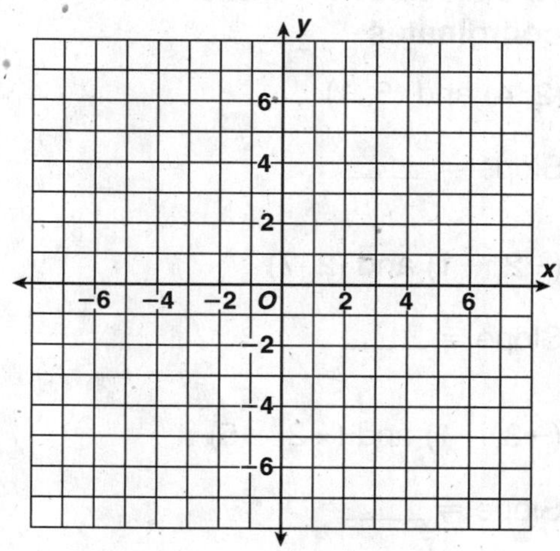

Name ______________________________ Date ___________ Class ___________

Focus on CAHSEE Standards: Intervention

STANDARD **1A6.0**

Skills Practice continued

Algebra 1 • Standard 6.0

Students graph a linear equation and compute the x- and y- intercepts (e.g., graph $2x + 6y = 4$). ~~They are also able to sketch the region defined by linear inequality (e.g., they sketch the region defined by $2x + 6y < 4$)~~.

Use the x- and y-intercepts to graph the following equation.

7. $y = 2x - 4$

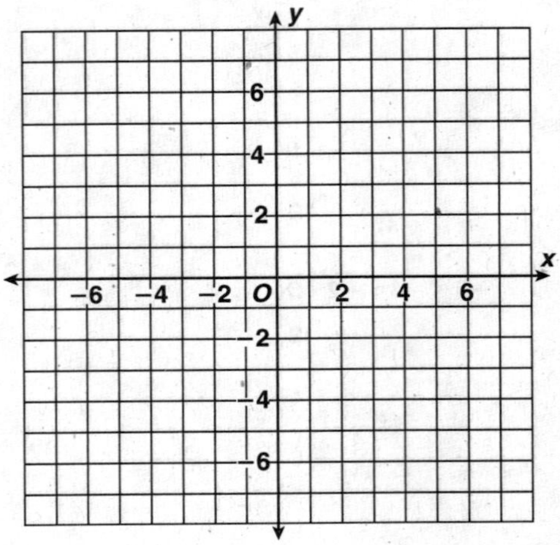

Find the slope of the lines below given two coordinates.

8. (2, 4) and (5, 1)

Slope = _____

9. (−2, −1) and (2, 7)

Slope = _____

10. (−3, −1) and (−2, −5)

Slope = _____

Use the slope-intercept form to graph the equations.

11. $y = 3 - x$

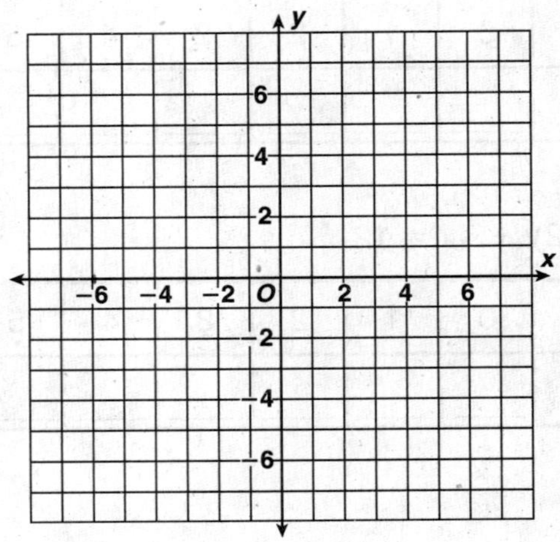

12. $y = 2x + 2$.

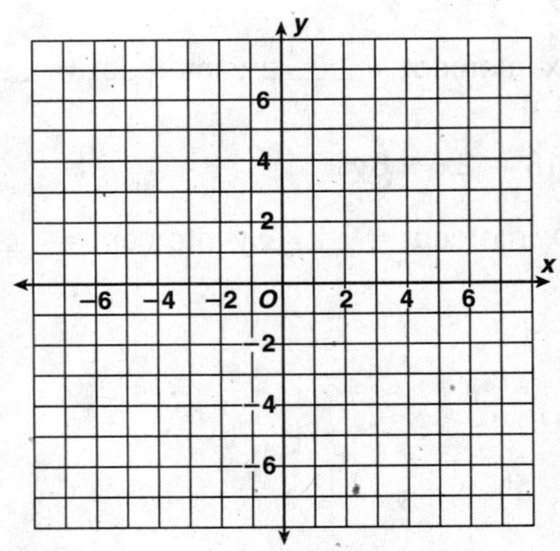

Name ______________________ Date __________ Class __________

STANDARD 1A7.0

Focus on CAHSEE Standards: Intervention

Alternative Strategies

Algebra 1 • Standard 7.0

Students verify that a point lies on a line, given an equation of the line. Students are able to derive linear equations by. ~~using the point-slope formula~~.

Use the standard form to write a linear equation.

1. Write $y = \frac{3}{2}x - 2$ in standard form.

Standard form is $Ax + By = C$.

$y = \frac{3}{2}x - 2$

First multiply both sides by 2 to clear the fraction.

$2(y) = 2$______________

Use the Distributive Property and multiply.

_____ $= 3x - 4$

Subtract $3x$ from both sides.

$2y - 3x =$ _____

In standard form.

____

2. Write $y = -\frac{1}{3}x + 2$ in standard form.

____

Use the slope-intercept form to write a linear equation.

3. Write the equation of the line with a slope of 2 and a y-intercept of -3.

The slope-intercept form of a line is $y = mx + b$, where $m =$ slope and $b = y$-intercept.

What values are you given for m and b?

$m =$ _____ $b =$ _____

Substitute for m and b in the equation $y = mx + b$.

____

4. Write the equation of the line with slope of $-\frac{1}{3}$ and y-intercept of 5.

____

Name ______________________________ Date ____________ Class ____________

STANDARD **1A7.0**

Focus on CAHSEE Standards: Intervention

Sample Problems

Algebra 1 • Standard 7.0

Students verify that a point lies on a line, given an equation of the line. Students are able to derive linear equations. ~~by using the point-slope formula~~.

1. Find the equation of the line that passes through the points (1, 2) and (7, 5).

When you are given two points, you can find the slope using $m = \frac{y_2 - y_1}{x_2 - x_1}$.

Substitute. For (x_1, y_1) use (1, 2) and for (x_2, y_2) use (7, 5).

$m = \frac{\_\_\_\_ - \_\_\_\_}{\_\_\_\_ - \_\_\_\_} = \_\_\_\_ = \_\_\_\_$

Use the slope-intercept form to find the equation of the line.

Choose one of the coordinate points to substitute into $y = mx + b$. You can use (7, 5) for x and y.

What does m represent in the above equation? ______________________

Substitute for m, x and y in the equation $y = mx + b$. Then solve for b.

$\_\_\_\_ = \_\_\_\_ \times \_\_\_\_ + b$

$= \_\_\_\_ + b$

$\_\_\_\_ = b$

Rewrite the equation substituting in the values for m and b.

$y = \_\_\_\_x + \_\_\_\_$

2. Find the equation of the line that has a slope of 2 and passes through the point (6, 3).

When you are given one point and the slope, you can use the slope-intercept form, $y = mx + b$, to find the equation of a line.

You already have values for x, y, and the slope, m. You need to identify b, the

____

Substitute the values for y, m, and x into the slope-intercept form and solve for b.

$\_\_\_\_ = \_\_\_\_ \bullet \_\_\_\_ + b$

$3 = \_\_\_\_ + b$ Multiply.

Subtract 12 from both sides of the equation.

$3 = 12 + b$

$-12 \quad -12$

$\_\_\_\_ = b$

Rewrite the equation substituting in the values for m and b.

$y = \_\_\_\_x + \_\_\_\_$

$y = \_\_\_\_\_\_\_\_$

Name ______________________ Date __________ Class __________

STANDARD 1A7.0

Focus on CAHSEE Standards: Intervention

Skills Practice

Algebra 1 • Standard 7.0

Students verify that a point lies on a line, given an equation of the line. Students are able to derive linear equations. ~~by using the point-slope formula~~.

Find the equation of the line given the y-intercept and the slope.

1. *y*-intercept = 3, slope = −6

$y =$ ______________________

2. $m = 2$, *y*-intercept = −1

$y =$ ______________________

3. *y*-intercept = −4, slope = $-\frac{8}{9}$

$y =$ ______________________

Find the equation of a line, in slope-intercept form, given two points.

4. (3, 2) and (5, 4)

$y =$ ______________________

5. (4, 8) and (3, 10)

$y =$ ______________________

6. (−3, −2) and (−7, −4)

$y =$ ______________________

Write the following linear equations in standard form.

7. $\frac{3}{4}x - 4 = y$

____

8. $y = \frac{7}{6}x + 3$

____

9. $y = 2x + 3\frac{3}{7}$

____

Write the equation of a line in slope-intercept form given a point and the slope.

10. Find the equation of the line that passes through (7, 3) and has a slope of −3.

____

11. Find the equation of the line that passes through (10, 2) and has a slope of $\frac{3}{5}$.

____

Name ______________________ Date ____________ Class ____________

Focus on CAHSEE Standards: Intervention

STANDARD 1A7.0

Problem Solving Practice

Algebra 1 • Standard 7.0

Students verify that a point lies on a line, given an equation of the line. Students are able to derive linear equations. ~~by using the point-slope formula.~~

Write a linear equation by using the information that is given in a table.

1. The table shows the cost of renting a video game, including a deposit.

Number of nights game is rented (x)	Cost of game rental (y)
1	$2
3	$4
7	$8

If nights (x) are graphed on the x-axis and cost (y) on the y-axis, what equation of a line fits the data?

Step 1: What are you asked to find?

____

Step 2: What are the facts?
What ordered pairs are you given?

____

Step 3: Break down the problem.
Use the slope-intercept form, $y = mx + b$. For x and y, choose a point. For the slope (m), substitute two points into the slope formula.

$$m = \frac{4 - 2}{\square - \square} = \frac{\square}{\square} = \_\_\_\_$$

Step 4: Solve the problem.
Substitute and solve for b. $y = mx + b$

____ = ____ • ____ + b

b = ____ − ____ = ____

What equation of a line fits the data?

____

Step 5: Check your answer makes sense.

2. Mr. Foster orders school shirts. The cost, with shipping, is in the table.

Number of shirts (x)	Cost of shirts (y)
2	$30
10	$120
50	$570

If the number of shirts (s) is graphed on the x-axis and cost (c) on the y-axis, what equation of a line fits the data?

Step 1: What are you asked to find?

____

Step 2: What are the facts?
What ordered pairs are you given?

____

Step 3: Break down the problem.
Use the slope-intercept form, $y = mx + b$. For x and y, choose a point. For the slope (m), substitute two points into the slope formula.

$$m = \frac{120 - 30}{\square - \square} = \frac{\square}{\square} = \_\_\_\_$$

Step 4: Solve the problem.
Substitute and solve for b. y = mx + b

____ = ____ • ____ + b

b = = ____

What equation of a line fits the data?

____

Step 5: Check your answer makes sense.

Name ______________________________ Date ____________ Class ____________

Focus on CAHSEE Standards: Intervention

STANDARD **1A8.0**

Sample Problems

Algebra 1 • Standard 8.0

Students understand the concepts of parallel lines ~~and perpendicular lines~~ and how those slopes are related. ~~Students are able to find the equation of a line perpendicular to a given line that passes through a given point~~.

1. What is the slope of a line parallel to the line $y = \frac{1}{2}x + 1$?

This line is in slope-intercept form.

What is the slope of the given line? ____

Are the slopes of parallel lines the same or different? ______________

What is the slope of a line parallel to the line $y = \frac{1}{2}x + 1$? ____

2. Write the equation of a line parallel to the line $y = 5x - 2$

The given line is in slope-intercept form.

What is the slope of the given line? ____

What is the y-intercept? ____

A line parallel to the given line will have the same slope and a different y-intercept.

Write an equation of a line parallel to $y = 5x - 2$:

$y =$ ____x ______________________

____

3. What is the slope of a line parallel to the line below?

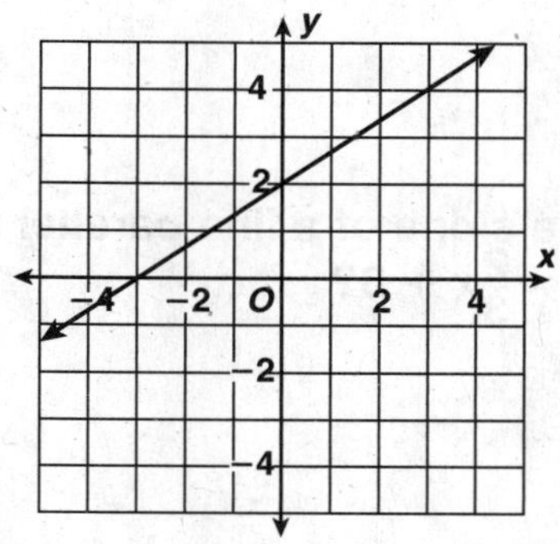

The slope of a given line is equal to rise/run, or the change in y-coordinates divided by the change in x-coordinates for any two points on the line.

Calculate the "rise," or change in y-coordinates between the points (0, **2**) and (3, **4**):

$4 - 2 =$ ____

Calculate the "run," or change in x-coordinates between the points (**0**, 2) and (**3**, 4):

$3 - 0 =$ ____

Calculate the slope, or $\frac{\text{rise}}{\text{run}}$: $\frac{2}{\square}$

The slope of a line parallel to the given line will be identical.

What is the slope of a line parallel to the line shown? ____

Name ______________________ Date __________ Class __________

Focus on CAHSEE Standards: Intervention

STANDARD 1A8.0

Skills Practice

Algebra 1 • Standard 8.0

Students understand the concepts of parallel lines ~~and perpendicular lines~~ and how those slopes are related. ~~Students are able to find the equation of a line perpendicular to a given line that passes through a given point.~~

1. What is the slope of a line parallel to the line $y = 3x - 4$?

____

2. What is the slope of a line parallel to the line $y = \frac{2}{3}x + 6$?

____

3. Write the equation of a line parallel to the line $y = 5x + 1$

____

____

____

4. Write the equation of a line parallel to the line $y = \frac{1}{4}x - 3$

____

____

____

5. What is the slope of a line parallel to the line below?

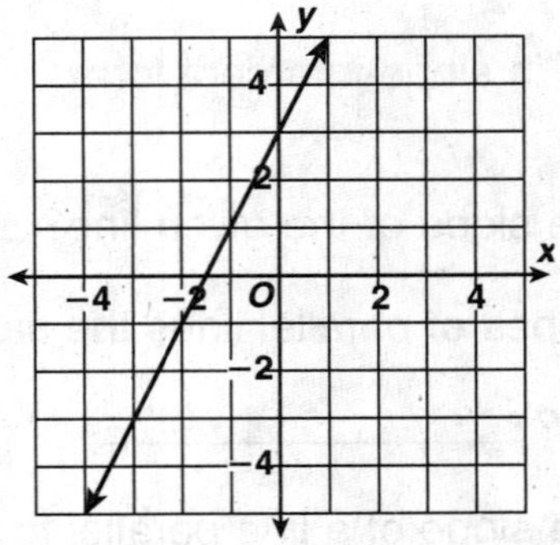

____

6. What is the slope of a line parallel to the line below?

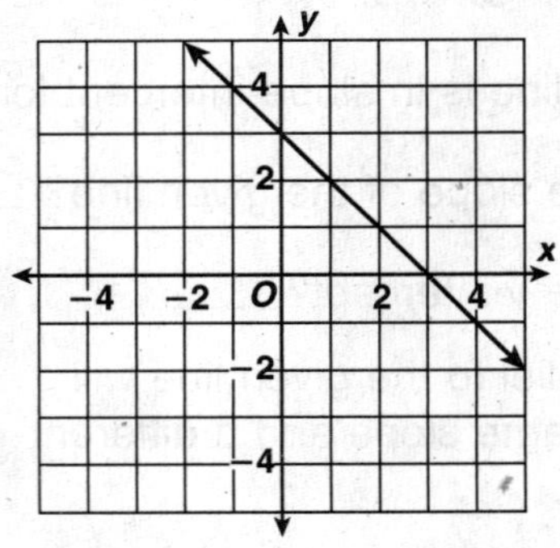

____

Name ______________________ Date __________ Class __________

STANDARD 1A9.0

Focus on CAHSEE Standards: Intervention

Alternative Strategies

Algebra 1 • Standard 9.0

Students solve a system of two linear equations in two variables algebraically and are able to interpret the answer graphically. Students are able to solve a system of two linear inequalities in two variables and to sketch the solution sets.

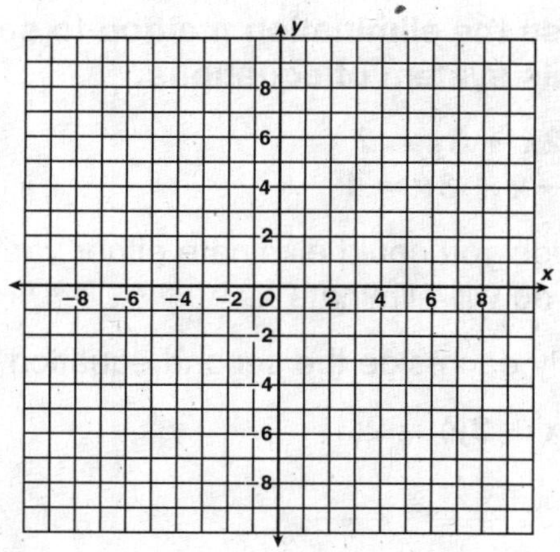

Use the coordinate plane to solve
$$\begin{cases} y = -x + 2 \\ y = x - 4. \end{cases}$$

1a. Make a table of values for $y = -x + 2$, using integer values of x from -3 to 4.

x	-3	-2	-1	0	1	2	3	4
y	5	4	___	2	___	___	___	___

1b. Graph $y = -x + 2$ on the coordinate plane.

1c. Make a table of values for $y = x - 4$, using integer values of x from -3 to 4.

x	-3	-2	-1	0	1	2	3	4
y	-7	___	___	-4	-3	___	___	___

1d. Graph $y = x - 4$ on the coordinate plane.

1e. At what point do the lines intersect?

____

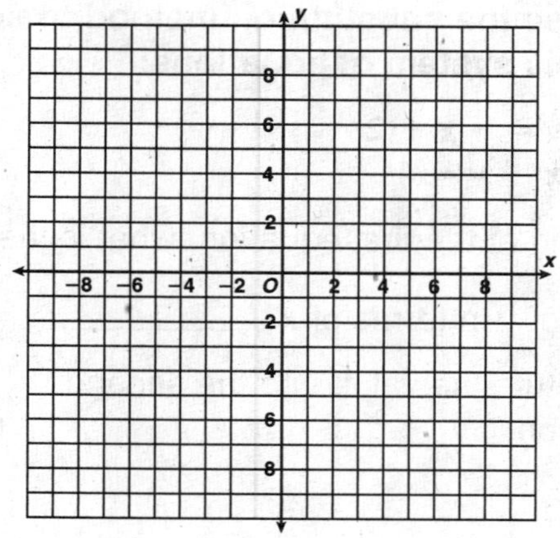

Use the coordinate plane to solve
$$\begin{cases} y \geq x + 2 \\ y \leq -x - 4. \end{cases}$$

2a. Graph $y = x + 2$.

2b. How is the graph of the linear equation $y = x + 2$ different from the graph of the linear inequality $y \geq x + 2$?

____

____

____

____

2c. Test a point such as (0, 0) by substituting into the inequality. __________

2d. Shade above the line.

2e. Graph the line $y = -x - 4$.

2f. Graph $y \leq -x - 4$.

2g. Which shaded region shows the solution of the system of inequalities?

____

Name ______________________ Date __________ Class __________

STANDARD 1A9.0

Focus on CAHSEE Standards: Intervention

Sample Problems

Algebra 1 • Standard 9.0

Students solve a system of two linear equations in two variables algebraically and are able to interpret the answer graphically. Students are able to solve a system of two linear inequalities in two variables and to sketch the solution sets.

1. Use the substitution method to solve this system of equations:

$$\begin{cases} y = -x + 2 \\ x - y = 4 \end{cases}$$

Looking at the first equation, what is the value of y in terms of x? __________

Substitute the value into the second equation.

$x - (\_\_\_\_\_\_\_\_\_\_) = 4$

Simplify and solve for x.

$2x - \_\_\_\_\_ = 4$

$2x = 6$

$x = \_\_\_\_\_$

Now substitute the value for x into the first equation.

$y = -\_\_\_\_\_ + 2$

Solve for y.

$y = \_\_\_\_\_$

Write the solution to the system of linear equations: (__________)

2. Use the elimination method to solve this system of equations:

$$\begin{cases} 2x + 3y = 7 \\ -x + 3y = 1 \end{cases}$$

Although you could eliminate either variable, here you will eliminate x.

Multiply each side the second equation by 2.

$2(-x + 3y) = 2(1)$

$-2x + \_\_\_\_\_ = \_\_\_\_\_$

Now add the result to the first equation.

$$\begin{array}{r} (2x + 3y = 7) \\ + \ (-2x + 6y = 2) \\ \hline \_\_\_\_\_ = 9 \end{array}$$

Solve for y.

$y = \_\_\_\_\_$

Now substitute the value for y into the first equation.

$2x + 3(\_\_\_\_\_) = 7$

Solve for x.

$x = 2$

Write the solution to the set of linear equations: (__________)

Name ______________________________ Date ____________ Class ____________

STANDARD 1A9.0

Focus on CAHSEE Standards: Intervention

Skills Practice

Algebra 1 • Standard 9.0

Students solve a system of two linear equations in two variables algebraically and are able to interpret the answer graphically. Students are able to solve a system of two linear inequalities in two variables and to sketch the solution sets.

Use the substitution method to solve the systems of equations.

1. $\begin{cases} y = 6x + 8 \\ x + y = -6 \end{cases}$

____

2. $\begin{cases} y = -2x + 5 \\ 3x + 2y = 6 \end{cases}$

____

Use the elimination method to solve the systems of equations.

3. $\begin{cases} 2x - 3y = 8 \\ x + 2y = -3 \end{cases}$

____

4. $\begin{cases} 4x + y = 10 \\ -x + 3y = -9 \end{cases}$

____

Name ______________________________ Date ____________ Class ____________

STANDARD 1A9.0

Focus on CAHSEE Standards: Intervention

Problem Solving Practice

Algebra 1 • Standard 9.0

Students solve a system of two linear equations in two variables algebraically and are able to interpret the answer graphically. Students are able to solve a system of two linear inequalities in two variables and to sketch the solution sets.

1. Gina plays center for her basketball team. In yesterday's game, she made 19 baskets and scored 28 points. Some baskets were worth 1 point and some were worth 2 points. How many 1-point shots, *x*, and 2-point shots, *y*, did Gina make?

Step 1: What are you asked to find? ________

____

____

Step 2: What are the facts?

How many baskets did Gina make? _____

How many points did she score? _____

Step 3: Break down the problem.
To find the number of 1-point shots, *x*, and 2-point shots, *y*, set up and solve a system of two equations.

$$\begin{cases} x + y = 19 \\ x + 2y = 28 \end{cases}$$

Step 4: Solve the problem.
Use elimination to solve the system of equations.

$x =$ _____ $y =$ _____

Gina made _____ 1-point shots and _____ 2-point shots.

Step 5: Check that your answer makes sense.
Substitute the solution values into each original equation to check that your solution is correct.

2. Tickets to the planetarium's laser show cost $3 for students and $4 for non-students. There were 34 tickets sold for a recent show, bringing in a total of $108. How many student tickets, *x*, and non-student tickets, *y*, were sold for the show?

Step 1: What are you asked to find? ________

____

____

Step 2: What are the facts?

How many tickets were sold in all? _____

How much money was brought in? ________

Step 3: Break down the problem.
To find the number of students tickets, *x*, and non-student tickets, *y*, set up and solve a system of two equations.

$$\begin{cases} x + y = 34 \\ 3x + 4y = 108 \end{cases}$$

Step 4: Solve the problem.
Use substitution to solve the system of equations.

$x =$ _____ $y =$ _____

_____ student tickets and _____ non-student tickets were sold for the laser show.

Step 5: Check that your answer makes sense.
Substitute the solution values into each original equation to check that your solution is correct.

Name ______________________ Date ____________ Class ____________

STANDARD 1A10.0

Focus on CAHSEE Standards: Intervention

Alternative Strategies

Algebra 1 • Standard 10.0

Students add, subtract, multiply, and divide monomials and polynomials. Students solve multistep problems, including word problems, by using these techniques.

Use algebra tiles to add polynomials.

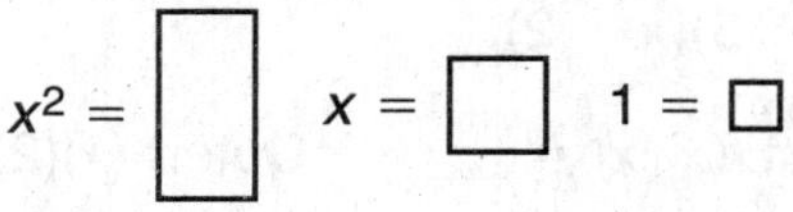

1a. Add $(3x^2 + 5x + 2)$ and $(2x^2 + x + 1)$.

1b. Model $(3x^2 + 5x + 2)$.

1c. Add the tiles for $(2x^2 + x + 1)$.

1d. How many x^2 tiles there are in all? _____

How many x tiles in all? _____

How many unit tiles? _____

1e. Write the sum in polynomial form.

____

Use algebra tiles to subtract polynomials.

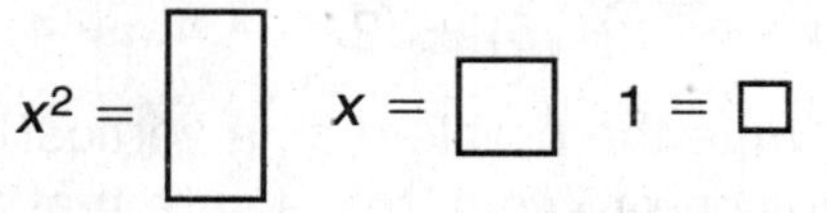

2a. Subtract $(2x^2 - 3x + 2)$ from $(4x^2 + 3x - 3)$.

2b. To subtract, use tiles to show the opposite of $2x^2 - 3x + 2$, which is $-2x^2 + 3x - 2$. Then add the polynomials.

2c. Model $(4x^2 + 3x - 3)$.

2d. Add the tiles for $-2x^2 + 3x - 2$.

2e. Write the sum in polynomial form.

____

Name ______________________ Date __________ Class __________

STANDARD 1A10.0

Focus on CAHSEE Standards: Intervention

Sample Problems

Algebra 1 • Standard 10.0

Students add, subtract, multiply, and divide monomials and polynomials. Students solve multistep problems, including word problems, by using these techniques.

Simplify using vertical addition.

1. $(3x^2 + 4x + 5) + (2x^2 + 4x + 4)$

   Rewrite the problem as a vertical addition problem. Make sure that similar terms are stacked above each other.

   $$\begin{array}{r} 3x^2 + 4x + 5 \\ + \; 2x^2 + 4x + 4 \\ \hline \end{array}$$

   Add the terms.

   $$\begin{array}{r} 3x^2 + 4x + 5 \\ + \; 2x^2 + 4x + 4 \\ \hline \end{array}$$

   $5x^2$ + ___ + ___

Simplify using the Distributive Property and vertical subtraction.

2. $(2x^2 + 2x + 2) - (2x^2 + x - 4)$

   Use the Distributive Property to turn the subtraction problem into an addition problem:
   $(2x^2 + 2x + 2) + (-2x^2 - x + 4)$

   Rewrite the problem as a vertical addition problem. Make sure that similar terms are stacked above each other.

   ____

   ____

   ______

Simplify by using the FOIL method.

3. $(x + 3)(x + 2)$

   First: $(x)(x)$ = ___ Outer: $(x)(2)$ = ___

   Inner: $(3)(x)$ = ___ Last: $(3)(2)$ = ___

   So: $(x + 3)(x + 2)$ =

   ___ + ___ + ___ + ___ =

   ____

Simplify by using the Distributive Property.

4. $-2x(6x^2 - x + 9)$

   Distribute $-2x$ across the parentheses:
   $-2x(6x^2 - x + 9)$

   $= -2x(6x^2)$ − ________ − ________

   = ______________________

Simplify.

5. $\dfrac{(14x^3 - 28x^2)}{7x}$

   What is the greatest common factor of each term? ____

   Factor using the GCF.

   $\dfrac{7x\,(2x^2 - 4x)}{7x}$

   $2x^2$ − ____

Name ______________________ Date ____________ Class ____________

STANDARD **1A10.0**

Focus on CAHSEE Standards: Intervention

Skills Practice

Algebra 1 • Standard 10.0

Students add, subtract, multiply, and divide monomials and polynomials. Students solve multistep problems, including word problems, by using these techniques.

1. How many terms does $2x^2 + x - 5$ have? ____________

2. What is the leading term in $3x^2 + 2x - 1$? ________

3. What degree is $7x^2 - 3x + 2$?

____

Add or subtract.

4. $(2x^2 + 3x) + (5x^2 - 2x - 9)$

____

5. $(5xy^2 - 2x^2y - 5xy - 7) + (-4xy^2 - 6xy + 3)$

____

6. $(7x^2 + 9x - 5) - (6x^2 - 10)$

____

7. $(23x^2y - xy^2 - 8) - (xy^2 - 8)$

____

Multiply.

8. $(x + 1)(x + 2)$

____

9. $(2x + 1)(x + 2)$

____

10. $x(x^2 + 3x - 2)$

____

11. $-x(2x^2 + x - 3)$

____

12. $4w^2(3w^3 - 2w^2 - w)$

____

Divide. (The denominator is not equal to zero.)

13. $\frac{(8x + 12)}{4}$

____

14. $\frac{6x^3 - 9x}{3x}$

____

Name ______________________ Date __________ Class __________

Focus on CAHSEE Standards: Intervention

STANDARD **1A10.0** *Problem Solving Practice*

Algebra 1 • Standard 10.0

Students add, subtract, multiply, and divide monomials and polynomials. Students solve multistep problems, including word problems, by using these techniques.

1. Five times a number minus 13 is 42. What is the number?

Step 1: What are you asked to find?

____

Step 2: What are the facts?

____

____

____

Step 3: Break down the problem.

Look closely at the language of the problem. "Five times a number" means an unknown number (x) that is multiplied by 5. So: $5x$. Also, "minus 13 is 42" means $5x - 13$ is equal to 42.

Write the equation. ______________

Step 4: Solve $5x - 13 = 42$.

$x =$ ____

Step 5: Check that your answer makes sense by putting the value for x into the equation.

$5(\_\_\_\_) - 13 = 42$

$\_\_\_\_ - 13 = 42$

$42 = 42$ ✓

Does the answer make sense? ________

2. Eight times the larger of two consecutive integers is 32 less than 9 times the smaller integer. What are the integers?

Step 1: What are you asked to find?

____

Step 2: What are the facts?

____

____

____

____

Step 3: Break down the problem.

Let a and $(a + 1)$ be the two consecutive integers.

Look closely at the language of the problem.

"Eight times the larger of two consecutive integers" means $8 \cdot (a + 1)$. The word "is" means "equals," so $8 \cdot (a + 1)$ equals a number that is 32 less than 9 times the smaller integer.

Restate the problem as an equation:

$8 \cdot (a + 1) =$ ______________

Step 4: Solve $8(a + 1) = 9a - 32$.

$a =$ ____ $(a + 1) =$ ____

What are the two consecutive integers?

____

Step 5: Check that your answer makes sense by putting the value for a into the equation. Does the answer make sense?

____

Name ________________________ Date ____________ Class ____________

STANDARD 1A15.0

Focus on CAHSEE Standards: Intervention

Alternative Strategies

Algebra 1 • Standard 15.0

Students apply algebraic techniques to solve rate problems, work problems, and percent mixture problems.

	Amount of Solution	Amount of Glycol
30% Glycol	x	
80% Glycol	$5 - x$	
60% Glycol	$x + (5 - x) =$ _____	

Use the table.

1. Mick is making a 5-quart antifreeze solution that contains 60% glycol. He is using solutions that contain 30% glycol and 80% glycol. How many quarts of each solution should he use to make the 60%-glycol solution?
   - **a.** Let x represent the amount of 30%-glycol solution. Use decimals to show the amount of glycol.
   - **b.** The amount of 80% glycol is the difference between 5 quarts and x.
   - **c.** The 60%-glycol solution combines the 30% and 80% solutions.
2. Solve the equation:
   $0.3x + 0.8(5 - x) = 0.6x + 0.6(5 - x)$
   - **a.** Use the Distributive Property.

     ____

     ____
   - **b.** Simplify ______________________
   - **c.** $x =$ _____
3. 30% solution: ______________________

   80% solution: ______________________

	r	t	$d = rt$
2 h	16		
2 h		2	

Use the table.

4. Brian took a 4-hour bicycle trip at an average rate of 18 mph. For the first 2 hours, he averaged 16 mph. How many miles per hour did he average for the last 2 hours of the trip?
   - **a.** Fill in the chart for rate, time, and distance = rate × time.
5. Use the information in the chart. Substitute into the equation below and then solve. (Hint: To find the total distance, add the two distances in the table.)
   - **a.** $\frac{\text{Total distance}}{\text{Total time}}$ = average rate for total trip

     $\frac{\square}{4} = 18$

     __________ = _____

     $2r =$ _____

     $r =$ _____
   - **b.** What is the rate for the last 2 hours?

     ____

Name ______________________________ Date ____________ Class ____________

STANDARD 1A15.0

Focus on CAHSEE Standards: Intervention

Sample Problems

Algebra 1 • Standard 15.0

Students apply algebraic techniques to solve rate problems, work problems, and percent mixture problems.

1. A lab is making a 30-kilogram metal alloy that is 30% copper out of an alloy that is 10% copper and an alloy that is 70% copper. How many kilograms of each alloy should they use to make the 30% copper alloy?

   a. Write a system of equations to solve the problem. First, identify the variables.

   Let $x =$ ____________________

   Let $y =$ ____________________

   b. Make the first equation equal to the number of kilograms of copper alloy.

   $x + y =$ ______

   c. Make the second set of equations equal to the percent of copper in each of the alloys.

   $0.1x + 0.7y =$ ______________

   d. Write the system of equations.

   ____

   ____

   e. You can solve the first equation for y and substitute for y in the second equation and solve. What is the equation you need to solve?

   ____

   e. 10% copper alloy: ______________

   70% copper alloy: ______________

2. Katrina can groom 3 dogs in 2 hours. Travis can groom 3 dogs in 3 hours. Working together, how many hours does it take Katrina and Travis to groom 3 dogs?

   a. When writing an equation for a work problem, you want to show the amount of work that each person can do in one hour. How much of the work can Katrina do in 1 hour? How much can Travis do in 1 hour?

   ____

   b. The amount of time required to complete the job is represented as $\frac{1}{t}$. Write the equation to represent this work situation.

   ______ + ______ $= \frac{1}{t}$

   c. Simplify and solve for t.

   ____

   ____

   d. How long does it take Katrina and Travis to groom 3 dogs if they work together?

   ____

   ____

Name ________________ Date ________ Class ________

STANDARD **1A15.0**

Focus on CAHSEE Standards: Intervention

Skills Practice

Algebra 1 • Standard 15.0

Students apply algebraic techniques to solve rate problems, work problems, and percent mixture problems.

Solve. Show your work.

1. Oceanside Stables offers scenic tours on horseback. A 4-hour tour averaged 3 miles per hour. If the average rate was 2.5 miles per hour for the first 3 hours, what was the average rate for the last hour?

   ____

2. Victor and Julia have a carpet-cleaning business. Victor can clean the carpets in 3 rooms in 2 hours. Julia can clean the carpets in 3 rooms in 4 hours. If they work together, how long will it take Victor and Julia to clean the carpets in 3 rooms?

   ____

3. A chemistry experiment requires 60 mL of a 40% alcohol solution. The lab has an 80% alcohol solution and a 30% alcohol solution. How many mL of the two solutions should be mixed to obtain 60 mL of a 40% alcohol solution? Write a system of equations you can use to solve the problem. Then solve.

   ____

   ____

4. Two cars leave St. Louis at the same time, one heading east and the other heading west. The eastbound car is traveling at an average speed of 64 miles per hour. The westbound car is traveling at an average speed of 52 miles per hour. After how much time will the cars be 550 miles apart?

   ____

Name ______________________ Date __________ Class __________

STANDARD 1A15.0

Focus on CAHSEE Standards: Intervention

Problem Solving Practice

Algebra 1 • Standard 15.0

Students apply algebraic techniques to solve rate problems, work problems, and percent mixture problems.

1. Jasmine can wash a car in 40 minutes on average. Nicholas can wash a car in 60 minutes. How long will it take them to wash a car together?

Step 1: What are you asked to find?

____

____

Step 2: What are the facts?

Jasmine can wash 1 car in _____ minutes, or $\frac{1}{40}$ of a car in 1 minute.

Nicholas can wash 1 car in _____ minutes. or

_____ of a car in 1 minute.

Step 3: Break down the problem.
Write an equation to represent the situation.

$$\frac{1}{\square} + \frac{1}{\square} = \frac{\square}{t}$$

What is the LCD? _________

Step 4: Solve the equation.
Multiply to get the common denominator.

$$\frac{\square}{120t} + \frac{\square}{120t} = \frac{\square}{120t}$$

$3t +$ _____ $=$ __________

_____ $= 120$

$t =$ _____

They can wash 1 car in _____ minutes.

Step 5: Check that your answer makes sense.

2. Gabrielle drove 5 hours to San Diego. Her average speed was 48 mph. She drove 42 miles per hour on average for the first 2 hours. What was her average speed for the final 3 hours of the trip?

Step 1: What are you asked to find?

____

____

Step 2: What are the facts?

The entire trip took _____ hours at an average speed of _____ mph. Gabrielle's average speed for the first _____ hours was _____ mph.

Step 3: Break down the problem.
Complete the table.

	r	t	$d = rt$
2 h		2	(2)
3 h	r		

Step 4: Solve the equation.
Use the table to write an equation and solve.

$$\frac{\square + \square r}{2 + \square} = 48$$

$$\frac{\square + \square r}{5} = 48$$

_____ $+ 3r =$ _____

$3r =$ _____

$r =$ _____

Gabrielle's average speed for the last 3 hours of the trip was _____ mph.

Step 5: Check your answer makes sense.

Name ______________________________ Date ____________ Class ____________

STANDARD 6PS1.1

Focus on CAHSEE Standards: Intervention

Vocabulary Strategies

Grade 6 • Statistics, Data Analysis, and Probability 1.1

KEY VOCABULARY WORDS	
Mean Example: The mean of the following data set is 3. 1, 1, 2, 2, 4, 4, 5, 5 $\frac{1+1+2+2+4+4+5+5}{8} = \frac{24}{8} = 3$	**Median** Example: The median of the following data set is 6. 1, 1, 6, 10, 10 **Mode** Example: The mode of the following data set is 32. 2, 3, 4, 5, 32, 32, 32

Use one of the key vocabulary words to complete each sentence.

1. The ____________________ of the following data set is 2.

   8, 6, 2, 2

2. The ____________________ of the following data set is 6.

   1, 1, 6, 19, 23

3. The ____________________ of the following data set is 6.

   1, 5, 6, 12

Name ______________________ Date ____________ Class ____________

STANDARD **6PS2.5**

Focus on CAHSEE Standards: Intervention

Vocabulary Strategies

Grade 6 • Statistics, Data Analysis, and Probability 2.5

KEY VOCABULARY WORDS

Conclusion
Example: After analyzing her data, Emma came to the conclusion that 90% of the students at Oakdale School would like to learn how to snowboard.

Graph
Example:

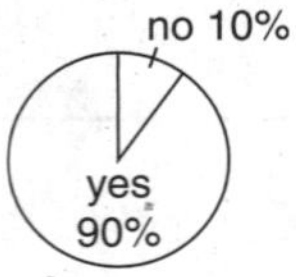

Invalid
Example: Dennis said Emma's conclusion was invalid because she asked a biased question.

Misleading
Example: Dennis said Emma's graph is misleading because it shows the results of an invalid survey.

Question
Emma's question was, "Would you like to learn how to snowboard?" But before she asked her question, she always said, "Snowboarding is really fun, and it's not that hard to learn how to do it."

Statement
Example: Dennis said that the statement Emma made before her question biases the question because it leads people to answer in a particular way.

Valid
Example: Dennis did his own survey and asked an unbiased question in order to come to a valid conclusion.

Use one of the key vocabulary words to complete each sentence.

1. A graph that displays data incorrectly is ______________.
2. When conducting a survey, it is important to ask an unbiased ______________.
3. If you ask a biased question in a survey, your conclusion will be ______________.
4. Gladys made a ______________ to display the data she gathered in her survey.
5. A sentence that ends in a period is a ______________.
6. If you ask an unbiased question to a sample that matches the population, your conclusion will be ______________.
7. You draw a ______________ from analyzing the results of the data in a survey.

Name ______________________ Date ____________ Class ____________

STANDARD **6PS3.1**

Focus on CAHSEE Standards: Intervention

Vocabulary Strategies

Grade 6 • Statistics, Data Analysis, and Probability 3.1

KEY VOCABULARY WORDS	
Combinations Example: Jessica discovered that she can wear 4 different combinations of outfits with 2 pairs of pants and 2 shirts.	**Theoretical Probability** Example: If Theresa rolls a number cube labeled 1 to 6, the theoretical probability of rolling a 1 is $\frac{1}{6}$.
Compound Event Example: Nikki established a compound event by first rolling a 7 on a number cube and then tossing a coin that landed heads up.	**Sample Space** Example: Richard puts 1 red marble, 1 blue marble, and 1 green marble in a jar. For his experiment, he picks 1 marble from the jar without looking. The sample space for the experiment is red, blue, and green.
Outcome Example: When Çarlos flips a coin, he knows that there are only 2 possible outcomes: heads or tails.	**Simple Event** Example: Marisa is trying to find the probability of the simple event of spinning an 8.

Use one of the key vocabulary words to complete each sentence.

1. Celeste spun a spinner and rolled a number cube. The ______________________ was 6 on the spinner and 4 on the number cube.

2. Melissa puts 1 pumpkin seed, 1 sunflower seed, and 1 cucumber seed in a pot. The ______________________ for her experiment is pumpkin seed, sunflower seed, cucumber seed.

3. The ______________________ that Mark will flip a coin and get tails is $\frac{1}{2}$, or 50%.

4. Elisa determined that the probability of the ______________________ of rolling 1 on a 6-sided number cube was 17%.

5. The possible ______________________ for spinning a spinner labeled 1–10 is one of these numbers: 1, 2, 3, 4, 5, 6, 7, 8, 9, 10.

6. The soccer team sold small and large T-shirts in both blue and red. There were 4 ______________________ of T-shirts to choose from.

Name ______________________________ Date ____________ Class ____________

STANDARD 6PS3.3

Focus on CAHSEE Standards: Intervention

Vocabulary Strategies

Grade 6 • Statistics, Data Analysis, and Probability 3.3

KEY VOCABULARY WORDS	
Outcome Example: The possible outcomes of flipping a coin are heads and tails. **Event** Example: When rolling a number cube, the event of rolling "an even number" consists of the outcomes 2, 4, and 6. **Probability** Example: A crate contains 2 red apples and 3 green apples. The probability of selecting a green apple is $\frac{3}{5}$.	**Theoretical Probability** Example: When you spin a spinner numbered 1–8, the theoretical probability of getting a 2 is $\frac{1}{8}$. **Reduce or Simplify** Example: If you reduce or simplify $\frac{6}{12}$, you get $\frac{1}{2}$. $\frac{6}{12} \div \frac{2}{2} = \frac{3}{6}$ $\frac{3}{6} \div \frac{3}{3} = \frac{1}{2}$

Use one of the key vocabulary words to complete each sentence.

1. If you ______________________ $\frac{5}{10}$, you will get $\frac{1}{2}$.
2. The possible ______________________ of spinning a spinner numbered 1–4 are 1, 2, 3, and 4.
3. The ______________________ that Mark will toss a coin and get tails is $\frac{1}{2}$, or 50%.
4. When spinning a spinner numbered 1–8, the ______________________ of spinning "a number greater than 5" consists of the outcomes 6, 7, and 8.
5. A pile of flash cards contains 7 math cards and 4 science cards. The ______________________ of selecting a math card is $\frac{7}{11}$.

Name ______________________________ Date ____________ Class ____________

Focus on CAHSEE Standards: Intervention

STANDARD 6PS3.5

Vocabulary Strategies

Grade 6 • Statistics, Data Analysis, and Probability 3.5

KEY VOCABULARY WORDS	
Dependent Events Example: A box has 2 green tennis balls and 4 orange tennis balls. To illustrate dependent events, Vaune selected a green ball and didn't replace it. Then she selected an orange ball from the box with the 5 remaining balls. **Independent Events** Example: A box has 2 green tennis balls and 4 orange tennis balls. To illustrate independent events, Vaune selected a green ball, put it back into the box, and then selected an orange ball.	**Probability** Example: A box has 2 green tennis balls and 5 orange tennis balls. The probability of selecting a green ball is $\frac{2}{7}$, and the probability of selecting an orange ball is $\frac{5}{7}$.

Use one of the key vocabulary words to complete each sentence.

A jar has 5 red counters and 7 blue counters.

1. The ____________________ of selecting a red counter is $\frac{5}{12}$.

2. Tori demonstrates she understands the meaning of ____________________ by picking a blue counter, replacing it, and then picking a red counter.

3. Aja demonstrates she understands the meaning of ____________________ by picking a red counter and then picking a blue counter without replacing the red one.

Name ______________________ Date ____________ Class ____________

Focus on CAHSEE Standards: Intervention

STANDARD 7NS1.1

Vocabulary Strategies

Grade 7 • Number Sense 1.1

KEY VOCABULARY WORDS	
Power of 10 Example: 10^4 is 10 to the fourth power and means $10 \times 10 \times 10 \times 10 = 10{,}000$.	**Scientific Notation** Example: 856 written in scientific notation looks like 8.56×10^2. **Standard Form** Example: 6.17×10^3 written in standard form is 6,170.

Use one of the key vocabulary words to complete each sentence.

1. 4,750 is the ______________ for 4.75×10^3.
2. 8.56×10^2 is ______________ for 856.
3. 10^3 is a number written using the ______________, and it means $10 \times 10 \times 10$, or 1,000.

Name ______________________ Date ____________ Class ____________

Focus on CAHSEE Standards: Intervention

STANDARD 7NS1.2

Vocabulary Strategies

Grade 7 • Number Sense 1.2

KEY VOCABULARY WORDS	
Common Denominator Example: $\frac{2}{5}$ and $\frac{3}{5}$ have a common denominator **Improper Fraction** Example: $\frac{11}{7}$ and $\frac{15}{8}$ are improper fractions **Mixed Number** Example: $1\frac{3}{5}$ and $2\frac{4}{7}$ are mixed numbers.	**Reciprocal** Example: $\frac{2}{3}$ is the reciprocal of $\frac{3}{2}$. **Simplest Form** Example: $\frac{4}{6}$ is in simplest form when simplified to $\frac{2}{3}$. **Simplify** Example: To simplify $\frac{4}{6}$, factor the numerator and denominator. $\frac{4}{6} = \frac{\cancel{2} \times 2}{\cancel{2} \times 3} = \frac{2}{3}$

Use one of the key vocabulary words to complete each sentence.

1. The ______________ of $\frac{5}{6}$ is $\frac{6}{5}$.
2. $\frac{3}{7}$ and $\frac{5}{7}$ have a ______________.
3. $2\frac{1}{7}$ is a ______________.
4. $\frac{22}{7}$ is a(n) ______________.
5. You can ______________ $\frac{5}{15}$ to $\frac{1}{3}$.
6. The fraction $\frac{1}{4}$ is in ______________.

Name ______________________ Date __________ Class __________

STANDARD 7NS1.3

Focus on CAHSEE Standards: Intervention

Vocabulary Strategies

Grade 7 • Number Sense 1.3

KEY VOCABULARY WORDS	
Decimal Example: A number that uses a decimal point to show tenths and hundredths, such as 5.42 **Fraction** Example: $\frac{4}{7}$ is a fraction.	**Percent** Example: A ratio comparing a number to 100 $45\% = \frac{45}{100}$

Use one of the key vocabulary words to complete each sentence.

1. 35%, which describes a ratio of 35 out of 100, is a __________.

2. The number 3.5 is a __________.

3. The number $\frac{3}{5}$ is a __________.

Name ______________________ Date __________ Class __________

STANDARD 7NS1.6

Focus on CAHSEE Standards: Intervention

Vocabulary Strategies

Grade 7 • Number Sense 1.6

KEY VOCABULARY WORDS	
Percent of Increase When one number is increased to another, find the percent of increase by using the percent of change formula. 25 is increased to 50. $50 - 25 = 25$ $\frac{25}{50} = 0.50 = 50\%$	**Percent of Decrease** When one number is decreased to another, find the percent of decrease by using the percent of change formula. 100 is decreased to 81 $100 - 81 = 19$ $\frac{19}{100} = 0.19 = 19\%$ **Percent of Change** is found by a formula that can be used for both percent of increase and decrease. $\text{percent of change} = \frac{\text{amount of change}}{\text{original amount}}$

Use one of the key vocabulary phrases to complete each sentence.

1. To find the percent of increase or decrease use the ______________ formula.

2. When changing from a smaller to a greater number, you can find the ______________.

3. When changing from a greater to a smaller number, you can find the ______________.

Name ______________________________ Date ____________ Class ____________

Focus on CAHSEE Standards: Intervention

STANDARD 7NS1.7

Vocabulary Strategies

Grade 7 • Number Sense 1.7

KEY VOCABULARY WORDS

Commission
Example: Adam is paid 5% commission on his sales of magazines. He earned $5.00 for his sales total of $100.

Compound Interest
Example: For the last 2 years, Theresa has been earning compound interest on both the $100 principal of her savings and the interest that the principal earns.

Discount
Example: Gina reduced the price of the book for a total discount of $5.00 from the retail price of $10.00.

Markup
Example: After Robert added the markup to the wholesale price, the shoes cost an additional $15.00.

Simple Interest
Example: Sara put $100 into an account with a simple interest rate of 5%. After two years, the account will have earned a simple interest of $10.00.
$I \times 100 \cdot 0.05 \cdot 2 = 10$

Tax
Example: The 6% sales tax added $0.60 to the total cost of the $10.00 sunglasses.

Tip
Example: The 15% tip that Melinda gave the waiter added $1.50 to the total cost of the $10.00 lunch.

Use one of the key vocabulary words to complete each sentence.

1. The ____________ of the football was $4.00 more than the wholesale price.
2. The 15% ____________ that Karen left for the waiter amounted to almost $6.00.
3. Sharon earned $340.75 ____________________ over two years on both the principal of her savings and the interest that the principal had earned in the meantime.
4. Ted earned a 10% ________________ on his total sales of talent show tickets.
5. The additional 8% ____________ Steven paid added another $6.40 to the cost of the cell phone.
6. After 1 year, Jennifer had earned $6.00 of ________________ on her $100 savings deposit.
7. Beth reduced the retail price of the vase for a total ____________ of $14.00.

Name ______________________ Date __________ Class __________

STANDARD 7NS2.1

Focus on CAHSEE Standards: Intervention

Vocabulary Strategies

Grade 7 • Number Sense 2.1

KEY VOCABULARY WORDS	
Base Example: $2^3 = 2 \cdot 2 \cdot 2 = 8$; 2 is the base. **Exponent** Example: $2^3 = 2 \cdot 2 \cdot 2 = 8$; 3 is the exponent.	**Expression** Example: $2^{-3} = \frac{1}{2^3} = \frac{1}{8}$ **Negative Exponent** A negative exponent, such as 2^{-3}, shows a value that is greater than zero but less than one. Example: $2^{-3} = \frac{1}{2^3}$

Use one of the key vocabulary words to complete each sentence.

1. $6x + 1$ is a mathematical ______________.

2. In the expression x^4, x is the ______________.

3. A value with a ______________ can be written as a fraction with a positive exponent in the denominator.

4. In the expression y^4, 4 is a(n) ______________.

Name ______________________________ Date ______________ Class ______________

Focus on CAHSEE Standards: Intervention

STANDARD 7NS2.2

Vocabulary Strategies

Grade 7 • Number Sense 2.2

KEY VOCABULARY WORDS	
Denominator Example: $\frac{3}{4}$ ← Denominator **Evaluate** Example: When you evaluate a numerical or algebraic expression, you find its value. **Greatest Common Divisior** Example: The greatest common divisor (GCD) of 10 and 15 is 5.	**Least Common Multiple** Example: The least common multiple (LCM) of 10 and 18 is 90. **Numerator** Example: $\frac{3}{4}$ ← Numerator **Simplify** Example: When you write a fraction in simplest form or lowest terms, you simplify it.

Use one of the key vocabulary words to complete each sentence.

1. The ______________________ of 15 and 20 is 60.
2. You can ______________________ both $\frac{5}{10}$ and $\frac{6}{12}$ to $\frac{1}{2}$.
3. The ______________________ in $\frac{1}{6}$ is 6.
4. The ______________________ of 27 and 36 is 9.
5. You need to know the values of x and y to ______________________ the expression $x + y$.
6. The fractions $\frac{2}{3}$ and $\frac{2}{5}$ have the same ______________________.

Name ______________________________ Date ______________ Class ______________

Focus on CAHSEE Standards: Intervention

STANDARD 7NS2.3

Vocabulary Strategies

Grade 7 • Number Sense 2.3

KEY VOCABULARY WORDS	
Exponent The number that indicates how many times the base is used as a factor. Example: $2^3 \cdot 2 \cdot 2 \cdot 2 = 8$; 3 is the exponent.	**Factor** A number that is multiplied by another number to get a product. Example: $2 \cdot 3 = 6$; 2 and 3 are factors.

Use one of the key vocabulary words to complete each sentence.

1. You can multiply one ______________ with another to get a product.
2. In n^3, 3 is the ______________.

Name ______________________ Date __________ Class __________

Focus on CAHSEE Standards: Intervention

STANDARD 7NS2.4

Vocabulary Strategies

Grade 7 • Number Sense 2.4

KEY VOCABULARY WORDS	
Base Example: $5^2 = 5 \cdot 5 = 25$; 5 is the base. **Exponent** Example: $5^2 = 5 \cdot 5 = 25$; 2 is the exponent.	**Perfect Square** Example: $5^2 = 25$, so 25 is a perfect square. **Square Root** Examples: $25 = 5 \cdot 5$ and $25 = -5 \cdot -5$, so 5 and -5 are square roots of 25.

Use one of the key vocabulary words to complete each sentence.

1. $8^2 = 64$, so 64 is a ____________.
2. In the expression 8^2, 8 is the ____________.
3. 10 and -10 are ____________ of 100.
4. In the expression x^5, 5 is the ____________.

Name ______________________ Date __________ Class __________

Focus on CAHSEE Standards: Intervention

STANDARD 7NS2.5

Vocabulary Strategies

Grade 7 • Number Sense 2.5

KEY VOCABULARY WORDS	
Absolute Value The absolute value of a number is the distance between that number and the zero on a number line. Examples: $\lvert 5 \rvert = 5$ and $\lvert -5 \rvert = 5$. **Absolute Value Sign** Absolute value signs are two vertical bars written in either side of a number expression. Examples: $\lvert 3 \rvert$, $\lvert -3 \rvert$	**Opposites** Two numbers that have the same absolute value but different signs are called opposites. Example: -5 and 5 are opposites.

Use one of the key vocabulary words to complete each sentence.

1. Since -15 and 15 have the same absolute value, they are ____________.
2. The vertical bars that surround -12 in $\lvert -12 \rvert$ are ____________.
3. The ____________ of $\lvert -4 \rvert$ and $\lvert 4 \rvert$ is 4.

Name ______________________ Date ____________ Class ____________

Focus on CAHSEE Standards: Intervention

STANDARD 7AF1.1

Vocabulary Strategies

Grade 7 • Algebra and Functions 1.1

KEY VOCABULARY WORDS	
Equation A mathematical sentence that shows that two expressions are equivalent. Example: $2m + 17 = 35$ **Expression** A mathematical phrase containing numbers, variables, and/or operations. Examples: $y + 5$, $2 + 18$	**Inequality** A statement that compares two expressions by using one of the following signs: $<$, $>$, $\leq$, $\geq$, $\neq$. Example: $x + 3 > 5$

Use one of the key vocabulary words to complete each sentence.

1. $24 - a$ is an ____________.
2. $24 - a = 5$ is an ____________.
3. $24 - a > 5$ is an ____________.

Name ______________________ Date ____________ Class ____________

Focus on CAHSEE Standards: Intervention

STANDARD 7AF1.2

Vocabulary Strategies

Grade 7 • Algebra and Functions 1.2

KEY VOCABULARY WORDS	
Expression A mathematical phrase that contains operations, numbers, and/or variables. Examples: 7×4, $3a + 5$, $12x$, $3(8 - y)$ **Order of Operations** A rule for evaluating expressions: first perform the operations in parentheses, then compute powers and roots, then perform all multiplication and division from left to right, and then perform all addition and subtraction from left to right.	**Variable** A symbol used to represent a quantity that can change. Example: In the expression $3x + 1$, x is the variable.

Use one of the key vocabulary words to complete each sentence.

1. $10 - 2 \times 4$ is equal to $10 - 8$ and not to 8×4 because of ____________________.
2. In $(3 + x) \div 2$, x is the ____________________.
3. $5y - 7$ is an example of a(n) ____________________.

Name ______________________ Date ____________ Class ____________

STANDARD 7AF1.5

Focus on CAHSEE Standards: Intervention

Vocabulary Strategies

Grade 7 • Algebra and Functions 1.5

KEY VOCABULARY WORDS

Bar graph
A graph that uses vertical or horizontal bars to display data.

Example:

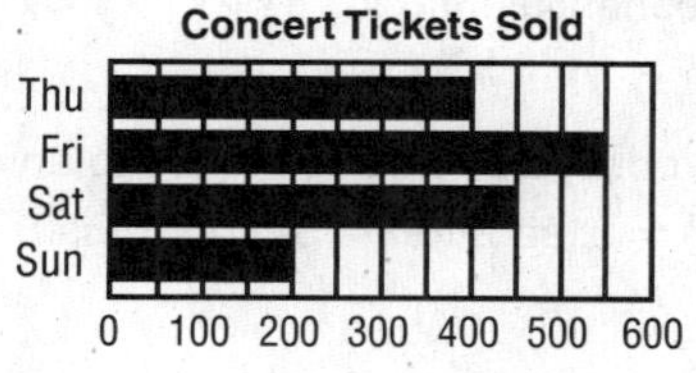

Circle graph
A graph that uses sections of a circle to compare parts to the whole and parts to other parts.

Example:

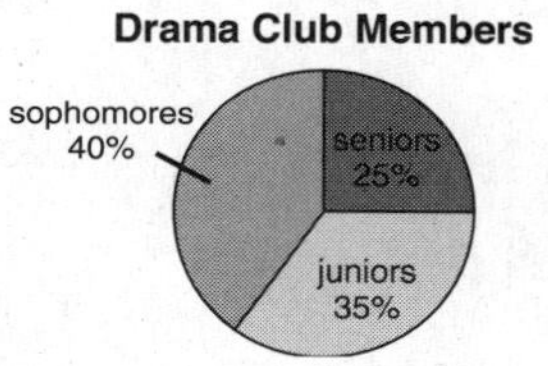

Venn diagram
A diagram that is used to show relationships between sets.

Example:

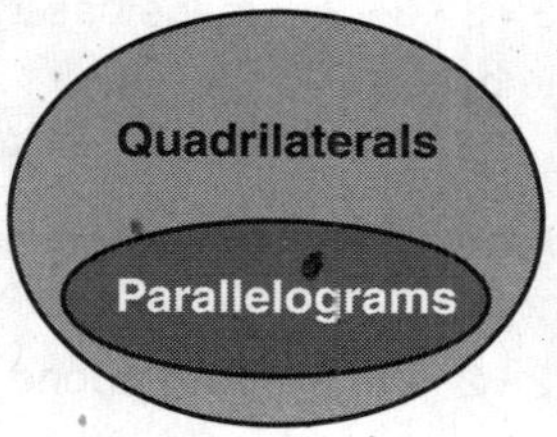

Use one of the key vocabulary words to complete each sentence.

1. Use a ____________________ to easily see to what degree one set of data overlaps with another.

2. Use a ____________________ to easily see how data changes from one measurement to the next.

3. Use a ____________________ to easily see how parts are related to each other and to a whole.

4. Use a ____________________ to easily see by height or length how data compare.

Name ______________________________ Date ____________ Class ____________

STANDARD **7AF2.1**

Focus on CAHSEE Standards: Intervention

Vocabulary Strategies

Grade 7 • Algebra and Functions 2.1

KEY VOCABULARY WORDS	
Base When a number is raised to a power, the number that is used as a factor is the base. Examples: $3^5 = 3 \cdot 3 \cdot 3 \cdot 3 \cdot 3$; 3 is the base. $x^3 = x \cdot x \cdot x$; x is the base.	**Exponential Form** A number is in exponential form when it is written with a base and an exponent. Examples: 4^2 is the exponential form for $4 \cdot 4$. y^5 is the exponential form for $y \cdot y \cdot y \cdot y \cdot y$.
Exponent The number that indicates how many times the base is used as a factor. Examples: $2^3 = 2 \cdot 2 \cdot 2 = 8$; 3 is the exponent. $y^4 = y \cdot y \cdot y \cdot y$; 4 is the exponent.	**Power** A number produced by raising a base to an exponent. Example: $2^3 = 8$, so 2 to the 3rd power is 8.

Use one of the key vocabulary words to complete each sentence.

1. In y^4, 4 is the __________.
2. In 3^2, 3 is the __________.
3. In $2^4 = 16$, 16 is the 4th __________ of 2.
4. x^3 is the ________________ for $x \cdot x \cdot x$.

Name ______________________________ Date ____________ Class ____________

STANDARD **7AF2.2**

Focus on CAHSEE Standards: Intervention

Vocabulary Strategies

Grade 7 • Algebra and Functions 2.2

KEY VOCABULARY WORDS	
Exponent The number that indicates how many times a base is used as a factor. Examples: $2^3 = 2 \cdot 2 \cdot 2 = 8$; 3 is the exponent. $y^4 = y \cdot y \cdot y \cdot y$; 4 is the exponent. **Monomial** A number or a product of numbers and variables with exponents that are whole numbers. Examples: $3x^2y^4$, a, b^8	**Square (numeration)** A term raised to the second power. Examples: In 5^2, the number 5 is squared. In x^2, the variable x is squared. **Square Root** One or more of the equal factors of a term. Examples: $16 = 4 \cdot 4$ and $16 = -4 \cdot -4$, so 4 and -4 are square roots of 16. $x^6 = x^3 \cdot x^3$, so x^3 is the square root of x^6.

Use one of the key vocabulary words to complete each sentence.

1. In y^2, the variable y is raised to the second power; so y^2 is the ______________ of y.
2. $x^{10} = (x^5)^2 = x^5 \cdot x^5$, so x^5 is ______________ of x^{10}.
3. In $2a^5$, the 5 is the ______________.
4. $2a^5$ is a ______________, the product of a number and a variable with a whole number exponent.

Name ______________________ Date __________ Class __________

STANDARD **7AF3.1**

Focus on CAHSEE Standards: Intervention

Vocabulary Strategies

Grade 7 • Algebra and Functions 3.1

KEY VOCABULARY WORDS	
Squared—multiplied by itself Example: 4 squared = $4^2 = 4 \times 4$ **Integer** A member of the set of whole numbers and their opposites. Examples: . . ., −3, −2, −1, 0, 1, 2, 3	**Cubed**—multiplied by itself twice Example: 5 cubed = $5^3 = 5 \times 5 \times 5$

Use one of the key vocabulary words to complete each sentence.

1. A number multiplied by itself two times is an example of a number that is __________.
2. A number that is multiplied by itself is an example of a number that is __________.
3. A negative whole number is an example of an __________.

Name ______________________ Date ____________ Class ____________

STANDARD 7AF3.3

Focus on CAHSEE Standards: Intervention

Vocabulary Strategies

Grade 7 • Algebra and Functions 3.3

KEY VOCABULARY WORDS	
Slope—a measure of the steepness of a line on a graph—the rise divided by the run. Example: The steepness of a skateboard ramp. **Negative Slope**—a line that slants downward Example: A car going down a hill. **Positive Slope**—a line that slants upward Example: A car going up a hill.	**Rise**—the vertical change when the slope of a line is expressed as the ratio $\frac{\text{rise}}{\text{run}}$ or "rise over run" Example: The rise between points (2, 3) and (6, 8) is 5. **Run**—the horizontal change when the slope of a line is expressed as the ratio $\frac{\text{rise}}{\text{run}}$ or "rise over run" Example: The run between points (2, 3) and (6, 8) is 4. **Ordered Pair**—a pair of numbers that can be used to locate a point on a coordinate plane Example: (3, −5)

Use one of the key vocabulary words to complete each sentence.

1. The ______________ is the difference between the *x* coordinates.
2. The ______________ is the difference between the *y* coordinates.
3. A ______________ goes downward from the left to the right.
4. A(n) ______________ is shown by an *x* and a *y* coordinate.
5. The steepness of a line is its ______________.
6. A ______________ goes upward from the left to the right.

Name ______________________ Date ____________ Class ____________

STANDARD 7AF3.4

Focus on CAHSEE Standards: Intervention

Vocabulary Strategies

Grade 7 • Algebra and Functions 3.4

KEY VOCABULARY WORDS

Ordered Pairs—a pair of numbers that can be used to locate a point on a coordinate grid.
Example: (3, 2) names a point on a coordinate grid.

Slope—a measure of steepness of a line on a graph; the rise divided by the run
Example: The line connecting points (0, 0) and (2, 4) has a slope of 2.

Ratio—a comparison of two quantities by division
Examples: 3 to 2, 3:2 or $\frac{3}{2}$.

Coordinate Grid—a plane formed by the intersection of a horizontal number line called the *x*-axis and a vertical number line called the *y*-axis.
Example: Any point on a coordinate grid can be named by an ordered pair.

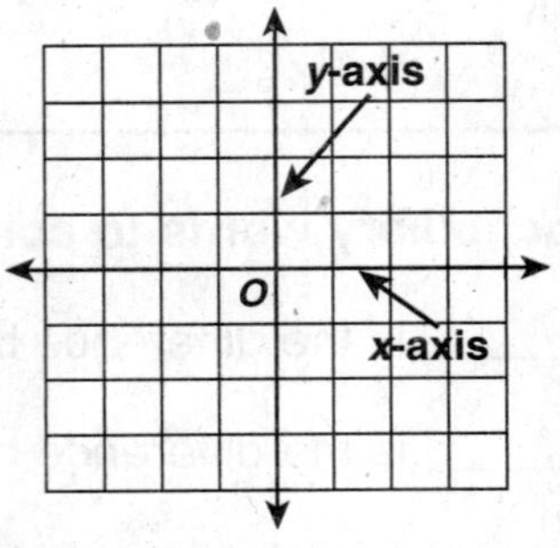

Use one of the key vocabulary words to complete each sentence.

1. An ______________ names a point on a coordinate grid.
2. To make a drink from a powdered mix, Alison can use the ______________ of mix to water.
3. To find the ______________ of a skateboard ramp, write a ratio for rise over run.
4. Ordered pairs can be displayed on a ______________.

Name ______________________________ Date ____________ Class ____________

STANDARD 7AF4.1

Focus on CAHSEE Standards: Intervention

Vocabulary Strategies

Grade 7 • Algebra and Functions 4.1

KEY VOCABULARY WORDS	
Equation a mathematical sentence that shows two expressions are equal Example: $6 - 2 = 4$; $3x + 10 = 16$ **Inequality** a mathematical sentence that shows two expressions are not equal Example: $7 < 9$; $4y - 3 > 8$ **Variable** a symbol used to represent a quantity that can change Example: x in the equation $5x + 6 = 21$.	**≥ (is greater than or equal to)** a symbol used to show that a quantity is equal to or greater than a second quantity Example: I need a score of 95% or better on the next test: $s \geq 0.95$. **≤ (is less than or equal to)** a symbol used to show that a quantity is equal to or less than a second quantity Example: I have to run the mile in a time of 10 minutes or less: $t \leq 10$.

Use one of the key vocabulary words to complete each sentence.

1. The solution to the ____________ $4y - 3 = 29$ is $y = 8$.
2. Mr. Crowe told his contractor to spend no more than \$2,500 on the new air conditioning unit: c ____ 2,500.
3. The voting age in the U.S. can be described by the ____________ $a \geq 18$.
4. To get a B in history, I need a score of 86% or better on the final, s ____ 86%.
5. To solve an equation or inequality, try to isolate the ____________.

Name ______________________________ Date ______________ Class ______________

STANDARD 7AF4.2

Focus on CAHSEE Standards: Intervention

Vocabulary Strategies

Grade 7 • Algebra and Functions 4.2

KEY VOCABULARY WORDS	
Miles per Hour (mi/h or mph) a customary unit for measuring speed Example: A plane that flies 3,000 miles in 6 hours has a speed of 500 mi/h. **Miles per Gallon (mi/gal)** a customary unit for measuring gas mileage Example: A car that goes 270 miles on 10 gallons of gas gets gas mileage of 27 mi/gal. **Ratio** a comparison of two quantities using division Example: There are 11 boys and 13 girls in Ms. Tran's class. The ratio of boys to girls is $\frac{11}{13}$ (also written 11 to 13, or 11 : 13).	**Unit Price** a unit rate used to compare prices Example: Five pounds of bananas that cost \$2.95 have a unit price of \$0.59 per pound. **Unit Rate** a rate in which the second quantity is 1 unit Example: If he made 24 necklaces in 8 hours, the unit rate is 3 necklaces per hour.

Use one of the key vocabulary words to complete each sentence.

1. Mr. Dill got a speeding ticket for driving 45 ____________________ in a school zone.
2. To decide if the 100-ounce bottle of detergent was a better buy than the 50-ounce bottle, Marc compared their ____________________.
3. To make the fruit punch Kim used a ____________________ of 3 parts orange juice to 1 part grape juice.
4. 2,000 calories a day is an example of a ____________________.
5. To conserve gas and save money, Kofi bought a car that gets 35 ____________________.

Name ______________________________ Date ____________ Class ____________

STANDARD **7MG1.1**

Focus on CAHSEE Standards: Intervention

Vocabulary Strategies

Grade 7 • Measurement and Geometry 1.1

KEY VOCABULARY WORDS	
Cross product—the product of numbers on the diagonal when comparing two ratios Example: $\frac{3}{4} \times \frac{6}{8}$ $4 \cdot 6 = 24$, $3 \cdot 8 = 24$ **Equivalent ratios**—ratios that have the same comparison Example: $\frac{20}{25}$ and $\frac{40}{50}$ are equivalent ratios because they both simplify to $\frac{4}{5}$. **Proportion**—an equation that states two ratios are equivalent Example: An equation such as $\frac{20}{25} = \frac{40}{50}$ is a proportion	**Rate**—a ratio that compares two quantities measured in different units Example: Pete ran at a rate of 3 miles in 18 minutes. **Ratio**—a comparison of two quantities by division Example: There were 12 students for each teacher in school; this ratio can be written three ways: $\frac{12}{1}$, 12 to 1, or 12:1. **Unit Conversion Factor**—a fraction used in unit conversion in which the numerator and denominator represent the same amount in different units Example: $\frac{3 \text{ ft}}{1 \text{yd}}$ or $\frac{1 \text{ yd}}{3 \text{ft}}$. The unit conversion factor states that 1 yard contains 3 feet. **Unit Rate**—a rate in which the second quantity in the comparison is 1 unit Example: $\frac{18 \text{ minutes}}{3 \text{ miles}} = \frac{6 \text{ minutes}}{1 \text{ mile}}$. The unit rate is 1 mile every 6 minutes.

Use one of the vocabulary words to complete each sentence.

1. A ______________ is a comparison of two products by division.
2. A ratio that compares two quantities measured in different units is a ______________.
3. You cross multiply the numerators and denominators of two fractions to get a ______________.
4. $\frac{2}{3} = \frac{4}{6}$ is a ______________.
5. The ________________________ used to change miles to yards is $\frac{1{,}760 \text{ yd}}{1 \text{ mi}} = \frac{1{,}760 \text{ yd}}{1{,}760 \text{ yd}} = 1$.
6. The ratios $\frac{16}{24}$ and $\frac{6}{9}$ are ________________________ because they both simplify to $\frac{2}{3}$.
7. Drew knit 8 pairs of socks in 4 hours, for a ____________ of 2 pairs of socks per hour.

Name ______________________________ Date ____________ Class ____________

Focus on CAHSEE Standards: Intervention

STANDARD 7MG1.2

Vocabulary Strategies

Grade 7 • Measurement and Geometry 1.2

KEY VOCABULARY WORDS	
Scale—the ratio between two sets of measurements Example: Joe wanted to paint a mural of the Mona Lisa, so he took measurements of Da Vinci's painting so that his mural would be to scale. **Scale drawing**—a drawing that uses a scale to make an object smaller than or larger than the real object Example: Tom made a scale drawing of the dog house he planned to build.	**Scale factor**—the ratio used to enlarge or reduce similar figures Example: Each inch on a road map is equivalent to 10 miles, so the scale factor is 1 in : 10 mi. **Scale model**—a proportional model of a three-dimensional object Example: Tom made a scale model of his dog house from his scale drawing.

Use one of the vocabulary words to complete each sentence.

1. The architect took careful measurements of the building site so that her drawing would be to ______________.
2. She used a ______________ of 1 : 100 when she drew up plans for the new building.
3. Her assistant constructed a three-dimensional ______________ of the building.
4. A copy of the blueprints were distributed to city officials so each person could look at a ______________ of the project.

Name ______________________ Date __________ Class __________

STANDARD 7MG1.3

Focus on CAHSEE Standards: Intervention

Vocabulary Strategies

Grade 7 • Measurement and Geometry 1.3

KEY VOCABULARY WORDS	
Density the density of a substance is the mass per one unit of volume Example: The density of a rock with a mass of 8 g and a volume of 4 cm^3 is 2 g per cm^3. **Person-Hours** the number of hours one person works to complete a job Example: It took Sam 7 person-hours to complete the task.	**Rate** a ratio that compares two quantities measured in different units Example: miles per hour or grams per cubic centimeter **Rate of Speed** the average distance an object travels in one unit of time Example: 7 miles per hour

Use one of the key vocabulary words to complete each sentence.

1. Person-days and miles per hour are examples of ____________.
2. You can find the ____________ of an object by dividing its mass by its volume.
3. If a turtle moves 1 mile in 2 hours, its ____________ is $\frac{1}{2}$ mile per hour.
4. If it takes 4 workers 10 hours to complete a job, then each worker worked

   2.5 ____________.

Name ______________________ Date ____________ Class ____________

Focus on CAHSEE Standards: Intervention

STANDARD 7MG2.1

Vocabulary Strategies

Grade 7 • Measurement and Geometry 2.1

KEY VOCABULARY WORDS

Cylinder—a three-dimensional figure with two parallel, circular bases connected by a curved surface

Example:

Diameter—a line that passes through the center of a circle with endpoints on the circle

Example:

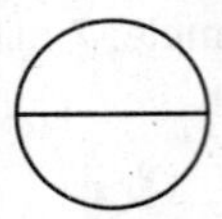

Perimeter—the distance around a polygon

Regular pentagon—a five-sided polygon with sides of equal length

Example:

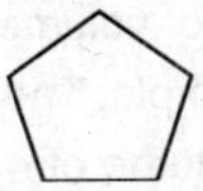

Surface area—the sum of the areas of the faces, or surfaces, of a polygon

Trapezoid—a quadrilateral with exactly one pair of parallel sides

Example:

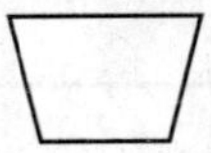

Use one of the key vocabulary words to complete each sentence.

1. The ______________ of a circle passes through its center.

2. If one side of a square is 5 cm long, then the ______________ of the square is 20 cm.

3. The formula for the ______________ of a rectangular prism is $2(\ell h + \ell w + wh)$.

4. While a ______________ has four sides, only two are parallel.

5. A can of soup is an example of a ______________.

6. If the perimeter of a ______________ is 10 cm, then each side of the figures measures 2 cm.

Name ________________________ Date ____________ Class ____________

STANDARD 7MG2.1

Focus on CAHSEE Standards: Intervention

Vocabulary Strategies

Grade 7 • Measurement and Geometry 2.2

KEY VOCABULARY WORDS

Area—the number of square units needed to cover a given surface.
Example: To find the area of a rectangle, multiply the length by the width.

Congruent figures—figures that have the same size and shape

Example:

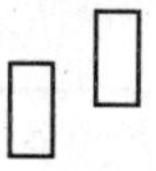

Octagon—a polygon with eight sides

Example:

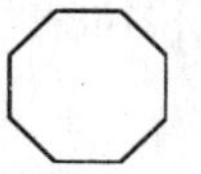

Right triangle—a triangle with exactly one right angle

Example:

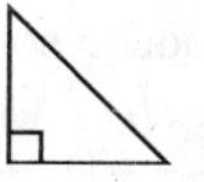

Use one of the key vocabulary words to complete each sentence.

1. A stop sign is in the shape of a(n) ____________.
2. If a square is cut along a diagonal line from the top left corner to the bottom right corner, two congruent ________________ form.
3. Two equilateral triangles with sides that measure 6 cm are ________________.
4. The product of a rectangle's length and width is its ____________.

Name ______________________ Date ____________ Class ____________

Focus on CAHSEE Standards: Intervention

STANDARD 7MG2.1

Vocabulary Strategies

Grade 7 • Measurement and Geometry 2.3

KEY VOCABULARY WORDS	
Face—a flat surface of a polyhedron. Example: 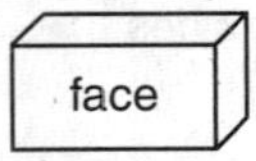	**Rectangular solid**—a polyhedron with rectangular bases and faces. Example: 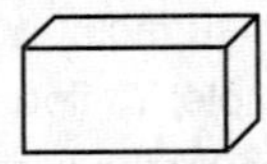
Surface area—the total area of all sides of a polyhedron. The formula for surface area of a rectangular solid is: $SA = 2(\ell \times w) + 2(w \times h) + 2(\ell \times h)$	**Volume**—the number of cubic units needed to fill a given space.

Use one of the key vocabulary words to complete each sentence.

1. To find the amount of paper needed to wrap a gift box, calculate the ____________ of the box.

2. A square and a rectangular prism both have 6 ____________.

3. A brick is in the shape of a ____________.

4. Since ____________ is the product of a rectangular solid's length, width, and height, it is expressed in cubic units.

Name ______________________ Date ____________ Class ____________

STANDARD 7MG2.3

Focus on CAHSEE Standards: Intervention

Vocabulary Strategies

Grade 7 • Measurement and Geometry 2.4

KEY VOCABULARY WORDS

Cubic foot—a cube that is one foot long, one foot wide, and one foot tall.

Example:

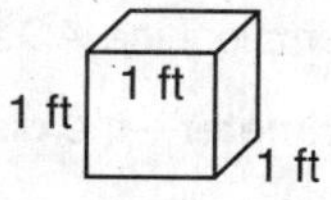

Cubic inch—a cube that is one inch long, one inch wide, and one inch tall.

Example:

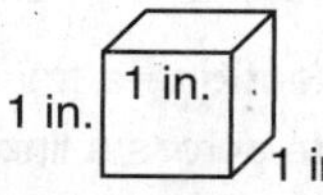

Square foot—a square with sides that measure 1 foot.

Example:

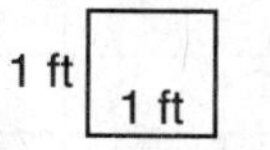

Square inch—a square with sides that measure 1 inch.

Example:

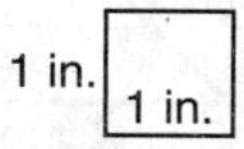

Use one of the key vocabulary words to complete each sentence.

1. One ____________ is approximately 16.38 cubic centimeters.
2. The customary unit best suited to measuring the area of a CD is the ____________.
3. One ____________ is equal to 144 square inches.
4. The customary unit best suited to measuring the volume of a sandbox is the ____________.

Name ______________________ Date __________ Class __________

STANDARD 7MG3.2

Focus on CAHSEE Standards: Intervention

Vocabulary Strategies

Grade 7 • Measurement and Geometry 3.2

KEY VOCABULARY WORDS

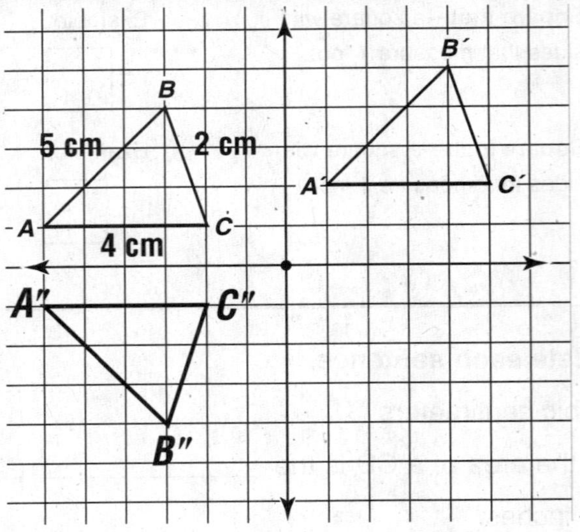

Coordinate—one of the numbers in an ordered pair that locates a point on a coordinate grid

Example: The coordinates of point *A* are (−6, 1).

Image—a figure resulting from a transformation

Example: The image of *A′B′C′* is *ABC* .

Parallel—lines that do not intersect in a plane.

Example: Line *AC* and line *A′C′* are parallel.

Perimeter—the distance around a polygon

Example: The perimeter of triangle *ABC* is 4 + 5 + 2 = 11 cm.

Reflection—a transformation of a figure that flips the figure across a line

Example: Triangle *A″B″C″* is the reflection of triangle *ABC* across the *x*-axis.

Translation—a movement or slide of a figure along a straight line

Example: Triangle *ABC* was translated 7 units right and 1 unit up to form the image triangle *A′B′C′*.

Vertex—(plural **Vertices**) point where two sides intersect

Example: In triangle *ABC*, the vertices are *A*, *B*, and *C*.

Use one of the key vocabulary words to complete each sentence.

1. A __________ of a figure flips the figure across a line.
2. The numbers that make up an ordered pair are called __________.
3. The figure that results from a translation or reflection is called a(n) __________.
4. The distance around a polygon is its __________.
5. A slide or movement along a straight line is a __________.
6. The point where the lines of a figure meet is called a __________.
7. Lines that do not intersect in a plane are __________.

Name ______________________ Date ____________ Class ____________

Focus on CAHSEE Standards: Intervention

STANDARD 7MG3.3

Vocabulary Strategies

Grade 7 • Measurement and Geometry 3.3

KEY VOCABULARY WORDS

Diagonal—$\overline{BD}$ is a diagonal of rectangle *ABCD*.

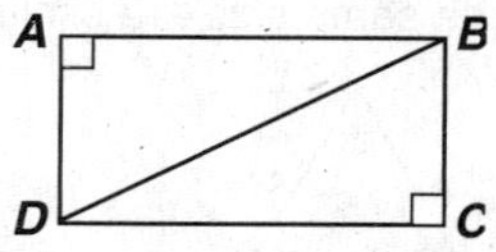

Hypotenuse—$\overline{BD}$ is a diagonal of rectangle *ABCD*. $\overline{BD}$ is also the hypotenuse of right triangle *ABD* and right triangle *DBC*.

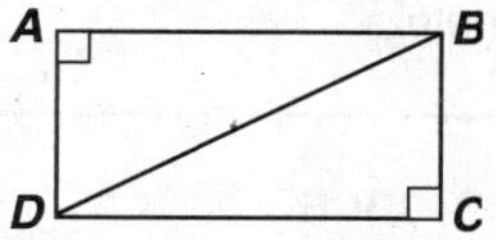

Line Segment—$\overline{AB}$ is a line segment connecting the points *A* and *B* on line *AB*

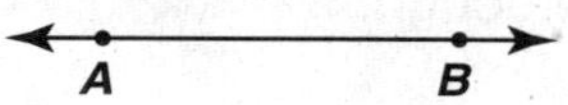

Multiples—6, 8, 10 is the second multiple of a 3, 4, 5 right triangle

Pythagorean theorem—In a right triangle, the area of the square on the hypotenuse is equal to the sum of the areas of the squares on the legs.

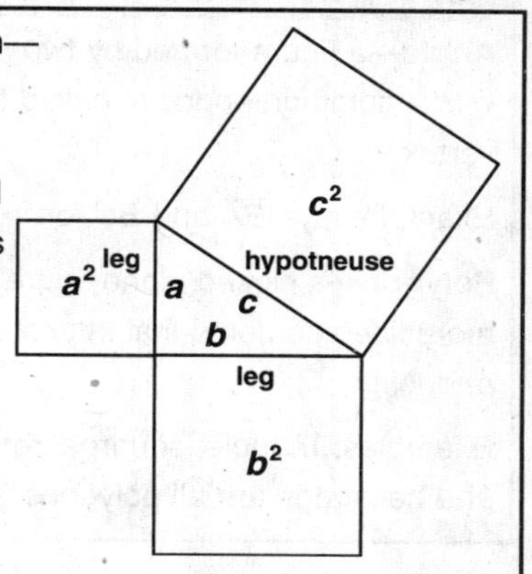

$a^2 + b^2 = c^2$

Right Triangle—The measure of angle *B* in right triangle *ABC* is 90°.

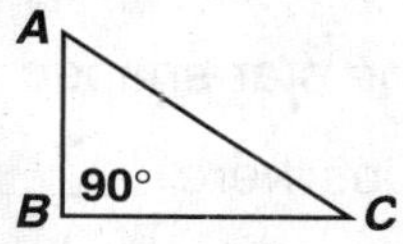

Use one of the key vocabulary words to complete each sentence.

1. A ______________ has an angle that measures 90°.
2. 9, 12, 15, are the third ______________ of 3, 4, 5.

Use this diagram for Exercises 3–5.

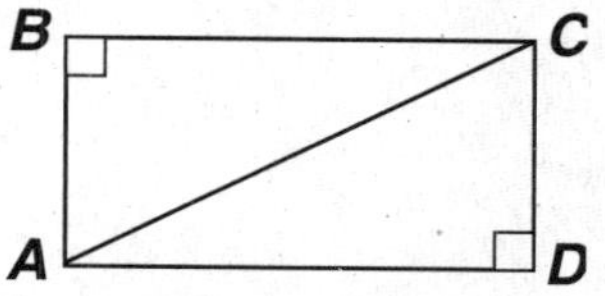

3. $\overline{AC}$ is a ______________ of rectangle *ABCD*.
4. $\overline{CD}$ is a ______________ connecting the points *C* and *D*.
5. $\overline{AC}$ is the ______________ of right triangle *ABC*.
6. The ______________ is used to show that in a right triangle, the sum of the squares of the legs is equal to the square of the hypotenuse or, $a^2 + b^2 = c^2$.

Name ______________________ Date ____________ Class ____________

STANDARD **7MG3.4**

Focus on CAHSEE Standards: Intervention

Vocabulary Strategies

Grade 7 • Measurement and Geometry 3.4

KEY VOCABULARY WORDS	
Angle—a figure formed by two rays with a common endpoint called the vertex Example: rays $\overrightarrow{BA}$ and $\overrightarrow{BC}$ form $\angle ABC$	**Similar**—Figures that are exactly the same shape but not necessarily the same size are similar. Example:
Polygon—a closed plane figure formed by three or more line segments that intersect only at their endpoints Examples: Triangle, squares, rectangles, pentagons, and hexagons are all polygons.	**Congruent**—having the same size and shape Example: Two squares are congruent only if their sides are the same length.

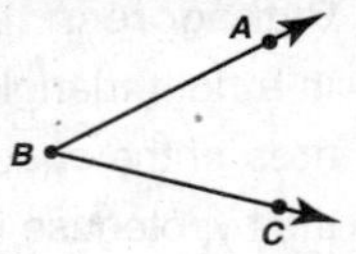

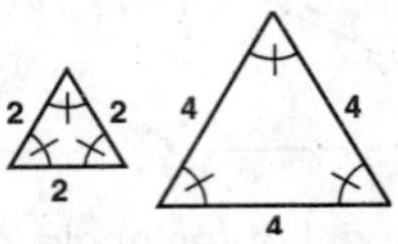

Use one of the key vocabulary words to complete each sentence.

1. At 3 P.M. the hands of a clock form a 90° ______________.
2. Kay used one star-shaped cutter to cut cookies from the dough.
   All her cookies were ______________.
3. The treehouse has 2 square windows, one 2 feet tall and one 3 feet tall.
   The windows are ______________.
4. A circle cannot be a ______________ because it is not composed of line segments.

Name ______________________ Date ____________ Class ____________

STANDARD 7PS1.1

Focus on CAHSEE Standards: Intervention

Vocabulary Strategies

Grade 7 • Statistics, Data Analysis and Probability 1.1

KEY VOCABULARY WORDS

Bar graph—a graph that uses vertical or horizontal bars to display data

Example:

Circle graph—a graph that uses sections of a circle to compare parts to the whole and parts to other parts

Example:

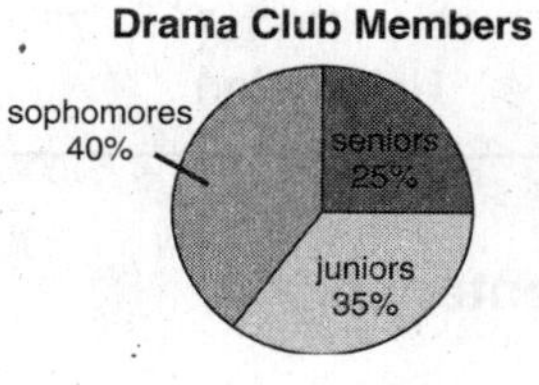

Data Set—set of information

Example: The runs scored by the Oakdale Green Sox

Line Graph—a graph that uses line segments to show how data changes

Example:

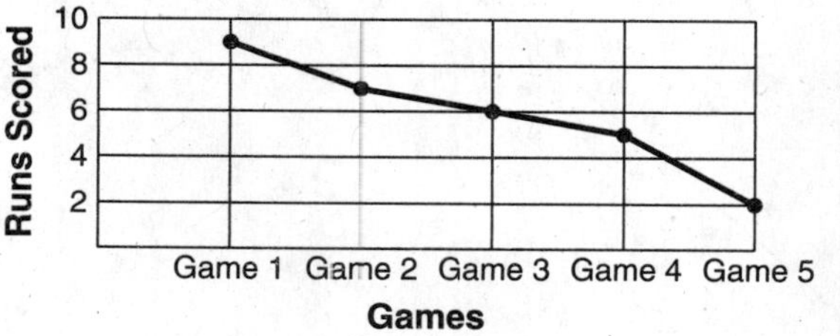

Line Plot

Example:

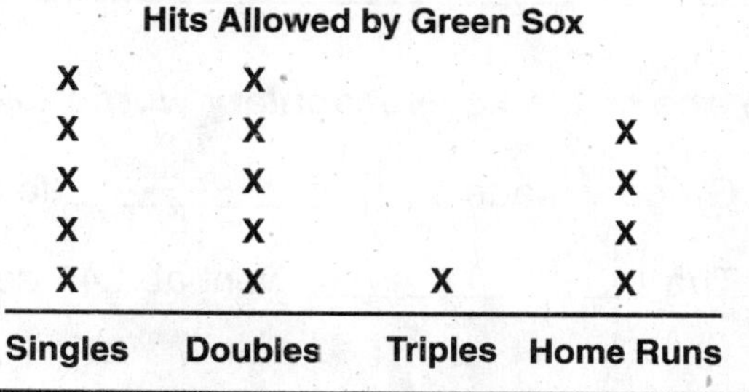

Use one of the key vocabulary words to complete each sentence.

1. A ______________ can help you show how something changes over several days.
2. A ______________ shows how data is distributed along a straight line.
3. Graphs, plots, and charts all display a ______________ of one kind or another.
4. A ______________ shows how parts are related to each other and to a whole.
5. A ______________ uses height or length to show how data compare.

Name ______________________________ Date ____________ Class ____________

STANDARD **7PS1.2**

Focus on CAHSEE Standards: Intervention

Vocabulary Strategies

Grade 7 • Statistics, Data Analysis and Probability 1.2

KEY VOCABULARY WORDS

Line of Best Fit—A straight line that comes closest to the points on a scatter plot

Example:

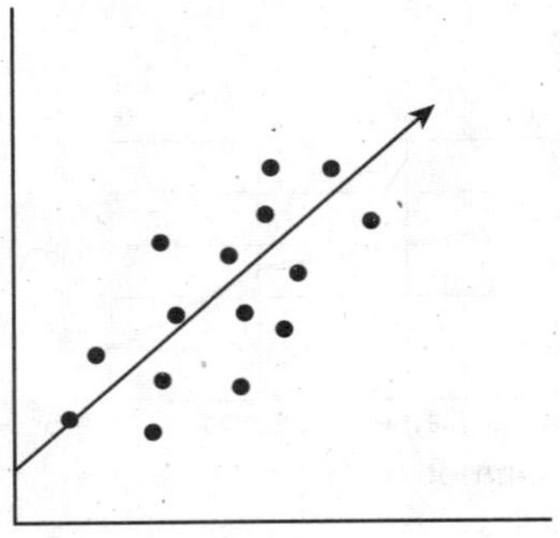

Scatter Plot

A graph with points plotted to show a possible relationship between two sets of data

Example:

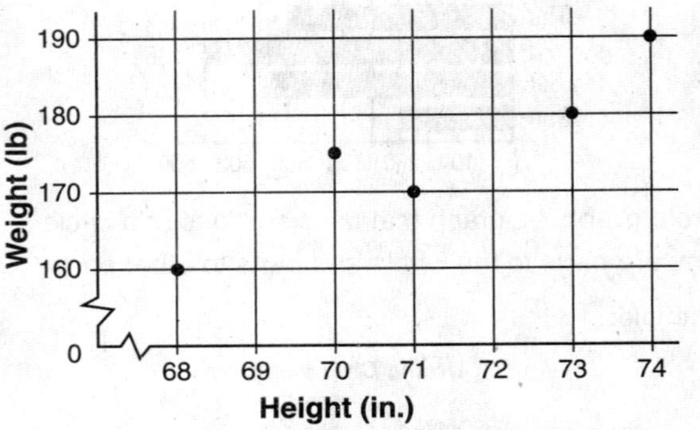

Use one of the key vocabulary words to complete each sentence.

1. Gregory made a ______________ to show how weight was related to height.

2. The ______________ went at a 45-degree angle from the top of the graph to the bottom.

Name ______________________ Date __________ Class __________

STANDARD 1A2.0

Focus on CAHSEE Standards: Intervention

Vocabulary Strategies

Algebra 1 • Standard 2.0

KEY VOCABULARY WORDS	
Radical A mathematical symbol indicating a positive root, or an expression containing such a symbol. Example: The radical expression $\sqrt{x+2}$, meaning "the positive square root of $x + 2$" contains a radical sign.	**Radicand** The expression under the radical sign. Example: In the radical expression $\sqrt{x+2}$, "$x + 2$" is the radicand.

Use one of the key vocabulary terms to complete each sentence.

1. A __________ sign indicates that you should take a root of the number or expression under that sign.

2. The __________ is the number or expression that is found under a radical sign.

Name ______________________ Date __________ Class __________

STANDARD 1A3.0

Focus on CAHSEE Standards: Intervention

Vocabulary Strategies

Algebra 1 • Standard 3.0

KEY VOCABULARY WORDS	
Absolute-Value Equation an equation that contains one or more absolute-value expressions Example: $\lvert x - 5 \rvert = 7$	**Absolute-Value Inequality** an inequality that contains one or more absolute-value expressions Example: $\lvert x - 5 \rvert > 7$

Use one of the key vocabulary words to complete each sentence.

1. $\lvert x + 2 \rvert \leq -2$ is an ______________________.

2. $\lvert x + 2 \rvert = -2$ is an ______________________.

Name ______________________ Date ____________ Class ____________

STANDARD **1A4.0**

Focus on CAHSEE Standards: Intervention

Vocabulary Strategies

Algebra 1 • Standard 4.0

KEY VOCABULARY WORDS	
Coefficient—a number multiplied by a variable Example: In $3x + 3$, the coefficient of x is 3. **Simplify**—perform all indicated operations Example: $-(20 - 13) + 8 =$ $-7 + 8 =$ 1	**Distributive Property of Multiplication**—for all real numbers a, b, and c, $a(b + c) = ab + ac$, and $(b + c)a = ba + ca$ Example: $3(x + 1) = 3x + 3$ **Like Terms**—terms with the same variables raised to the same exponents Example: $2x$ and $3x$ are like terms.

Use one of the key vocabulary words to complete each sentence.

1. If you ______________________ $6x + 3 + 2(x + 2)$, the result is $8x + 7$.

2. The expression $6x + 2 + x$ can be simplified to $7x + 2$ by combining ______________________.

3. In the expression $4t - 1$, the ______________________ of t is 4.

4. The ______________________ allows us to rewrite $4(y - 2)$ as $4y - 8$.

Name ______________________ Date __________ Class __________

STANDARD **1A5.0**

Focus on CAHSEE Standards: Intervention

Vocabulary Strategies

Algebra 1 • Standard 5.0

KEY VOCABULARY WORDS	
Inverse operation—an operation that "undoes" the result of another operation Example: In the equation $x + 7 = 9$, you can undo the operation of "+ 7" by subtracting 7 from both sides. **Linear equation**—an equation whose graph is a straight line Example: $5x + 3 = 12$	**Reciprocal**—The number $\frac{1}{a}$ is called the reciprocal of *a*. Example: $\frac{1}{4} \cdot 4 = 1$, so $\frac{1}{4}$ and 4 are reciprocals. **Equivalent equations**—equations that have the same solution Example: For both $x - 5 = 10$ and $x + 5 = 20$, $x = 15$.

Use one of the key vocabulary words to complete each sentence.

1. The graph of $3x - 4 = 8$ is a line. $3x - 4 = 8$ is a ____________________.

2. You can use the ____________________ to undo an operation when solving an algebraic equation.

3. Since $10 \cdot \frac{1}{10} = 1$, 10 and $\frac{1}{10}$ are ____________________.

4. ____________________, such as $x = 8$ and $x - 2 = 6$, have the same solution.

Name ______________________ Date __________ Class __________

STANDARD **1A6.0**

Focus on CAHSEE Standards: Intervention

Vocabulary Strategies

Algebra 1 • Standard 6.0

KEY VOCABULARY WORDS

Linear equation—an equation whose graph is a line

Example: $y = x - 1$

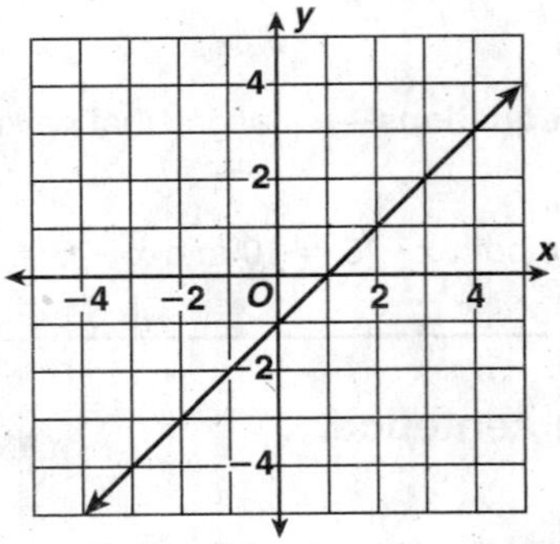

Input/output table—table used to graph an equation

Example: for the equation and graph above:

x	−1	0	1
y	−2	−1	0

Slope

Example: $m = \frac{y_2 - y_1}{x_2 - x_1} = \frac{-1 - 2}{-2 - 2} = \frac{3}{4}$

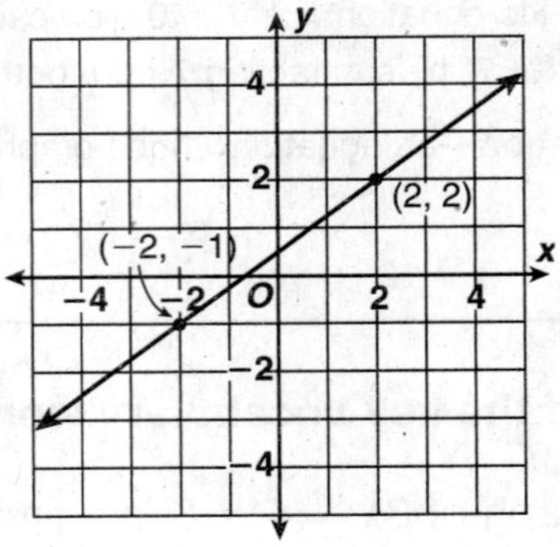

Slope-intercept form

Example: In the equation $y = -2x + 4$, the slope is −2 and the y-intercept is 4.

***x*-intercept**—where the graph of a linear equation crosses the x-axis ($y = 0$)

Example: For the equation $y = x - 1$, the x-intercept is 1. (See the graph at left.)

***y*-intercept**—where the graph of a linear equation crosses the y-axis ($x = 0$)

Example: For the equation $y = x - 1$, the y-intercept is −1. (See the graph at left.)

Use one of the key vocabulary words to complete each sentence.

1. The ______________________ is the point on a graph at which the value of y is zero.
2. The ______________________ of a line indicates how steep a line on a graph is.
3. $y = 3x$ is a ______________________.
4. The ______________________ is when $y = mx + b$, where m is the slope of the line and b is the y-intercept.
5. The point on a graph at which the x value is zero is the ______________________.
6. To show the values of y at several different values of x you could use

   a(n) ______________________.

Name ______________________ Date ____________ Class ____________

STANDARD **1A7.0**

Focus on CAHSEE Standards: Intervention

Vocabulary Strategies

Algebra 1 • Standard 7.0

KEY VOCABULARY WORDS	
Slope formula—an equation that is used to find the slope of a line when two points of the line are known Example: the line that passes through (2, 3) and (3, 4) has a slope of $m = \frac{y_2 - y_1}{x_2 - x_1} = \frac{4 - 3}{3 - 2} = \frac{1}{1} = 1$ **Rise**—the difference in the *y*-values of two points on a line Example: For the points (3, −1) and (6, 5), the rise is $5 - (-1) = 6$. **Run**—the difference in the *x*-values of two points on a line Example: For the points (3, −1) and (6, 5), the run is $6 - 3 = 3$. **Slope-intercept form**—an equation in the form $y = mx + b$, where *m* is the slope and b is the *y*-intercept Example: In $y = 4x + 1$, the slope is 4 and the *y*-intercept is 1.	**Standard form**—$Ax + By = C$, where *A*, *B*, and *C* are real numbers Example: $5x + 6y = 34$ ***y*-intercept**—the *y*-coordinate(s) of the points where a graph intersects the y-axis Example: The *y*-intercept is 2. 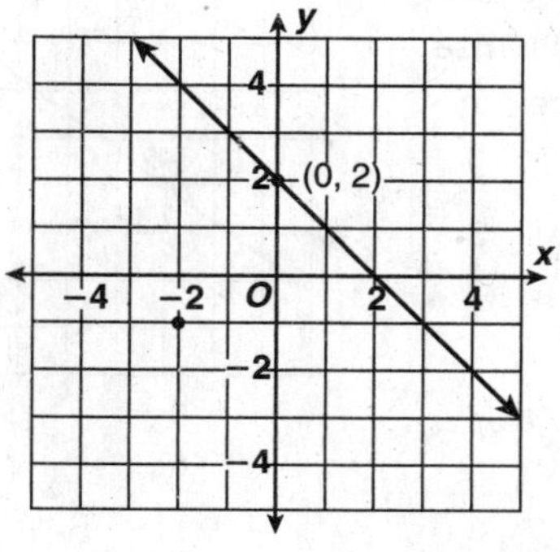

Use one of the key vocabulary words to complete each sentence.

1. In the equation $y = -2x + 3$, the ______________________ is 3.

2. The ______________________ can be used to find the slope of a line when you know two points on the line.

3. The difference between the *y*-values of two points on a line is called the ______________________.

4. The difference between the *x*-values of two points on a line is called the ______________________.

5. The equation $-2x + y = 3$ is in ______________________.

6. The equation $y = -2x + 3$ is in ______________________.

Name ______________________ Date ____________ Class ____________

STANDARD **1A8.0**

Focus on CAHSEE Standards: Intervention

Vocabulary Strategies

Algebra 1 • Standard 8.0

KEY VOCABULARY WORDS

Parallel lines

Lines in a plane that do not intersect.

Example:

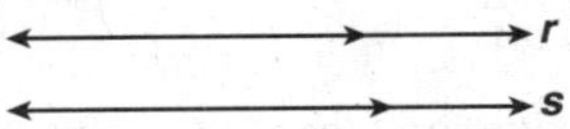

Slope

A measure of the steepness of a line on a graph; the rise divided by the run.

Example:

Slope $= \frac{\text{rise}}{\text{run}} = \frac{3}{4}$

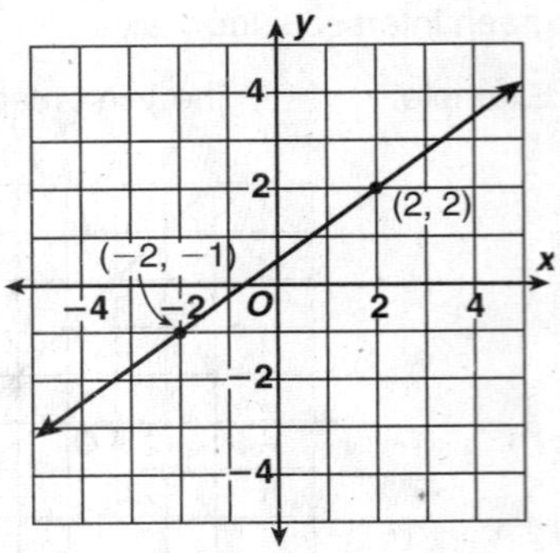

Slope-intercept form

A linear equation written in the form $y = mx + b$, where m represents slope and b represents the y-intercept.

Example: $y = 6x - 3$

***y*-intercept**

The y-coordinate of the point where the graph of a line crosses the y-axis.

Example:

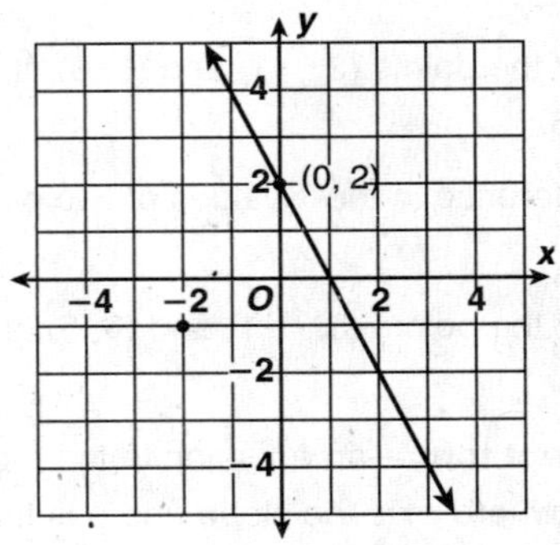

Use one of the key vocabulary words to complete each sentence.

1. In the line $y = 2x + 4$, 4 represents the ____________________.

2. In the line $y = 3x - 5$, 3 represents the ____________________ of the line.

3. The line $y = 4x + 1$ is written in the form $y = mx + b$, also known as ____________________.

4. The line $y = \frac{1}{3}x + 2$ is ____________________ to the line $y = \frac{1}{3}x + 5$.

Name ______________________ Date __________ Class __________

STANDARD 1A9.0

Focus on CAHSEE Standards: Intervention

Vocabulary Strategies

Algebra 1 • Standard 9.0

KEY VOCABULARY WORDS

Elimination method—a method for solving systems of equations that involves adding together individual equations in order to eliminate one variable and solve for a second variable

Example: $\begin{cases} x - y = 2 \\ 2x + y = 4 \end{cases}$

$$\begin{array}{r} x - y = 2 \\ + \; 2x + y = 4 \\ \hline 3x = 6 \\ x = 2 \end{array}$$

Linear equation—an equation whose solutions form a straight line on a coordinate plane

Example: $y = 2x + 1$

Linear inequality—a mathematical sentence using $<$, $>$, $\leq$, or $\geq$ whose graph is a region with a straight-line boundary

Example: $y \leq 2x + 1$

Substitution method—a method for solving systems of equations that involves isolating a variable in one equation and substituting the value into a second equation in order to solve for a second variable.

Example: $\begin{cases} x - y = 2 \\ 2x + y = 4 \end{cases}$

$$\begin{aligned} x &= y + 2 \\ 2(y + 2) + y &= 4 \\ 2y + 4 + y &= 4 \\ 3y &= 0 \\ y &= 0 \end{aligned}$$

System of linear equations—two or more linear equations in two or more variables

Example: $\begin{cases} x + y = -1 \\ -x + y = -3 \end{cases}$

System of linear inequalities—two or more linear inequalities in two or more variables.

Example: $\begin{cases} x + y \leq -1 \\ -x + y \geq -3 \end{cases}$

Use one of the key vocabulary words to complete each sentence.

1. The ______________________ of solving a system of equations involves isolating a variable in one equation and plugging the variable value into a second equation to solve for a second variable.

2. The ______________________ of solving a system of equations involves adding together two equations in order to reduce the number of variables to help solve for a single variable.

3. The solutions to a(n) ______________________ form a straight line on a coordinate plane.

4. The solutions to a(n) ______________________ form a region bounded by a straight line on a coordinate plane.

5. $\begin{cases} 2x + 3y \leq 6 \\ x + y \geq 2 \end{cases}$ is an example of a(n) ______________________.

6. $\begin{cases} 2x + 3y = 6 \\ x + y = 2 \end{cases}$ is an example of a(n) ______________________.

Name ______________________________ Date ____________ Class ____________

STANDARD 1A10.0

Focus on CAHSEE Standards: Intervention

Vocabulary Strategies

Algebra 1 • Standard 10.0

KEY VOCABULARY WORDS	
Binomial—a polynomial with two terms Examples: $3x + 2y$, $x^2 - 4x$ **Monomial**—a polynomial with one term Examples: $3x$, $2x^2$	**Polynomial**—one monomial or the sum or difference of monomials Examples: $3x$, $3x + 2y$, $9x^2 + 3x + 2$ **Trinomial**—a polynomial with three terms Examples: $9x^2 + 3x + 2$, $x^2 - 4x - 7$

Use one of the key vocabulary words to complete each sentence.

1. Monomials, binomials, and trinomials are all different kinds of ______________.

2. The expression $5a - 10$ is a ______________ because it has two terms.

3. The expression $5a - 10a + 6$ is a ______________ because it has three terms.

4. The expression $5a$ is a ______________ because it has one term.

Name ______________________ Date ____________ Class ____________

Focus on CAHSEE Standards: Intervention

STANDARD 1A15.0

Vocabulary Strategies

Algebra 1 • Standard 15.0

KEY VOCABULARY WORDS	
Percent Mixture Problem—a problem in which two or more solutions or substances are mixed to create a third solution or substance. Example: How many quarts of 30% antifreeze and 60% antifreeze should be mixed to produce 3 quarts of 50% antifreeze? **Work Problem**—a problem in which two or more individuals complete a job. The time it takes each individual to complete the job is known. Example: Ana can assemble 2 computers in 6 hours. Ian can assemble 2 computers in 7.5 hours. If they work together, how many hours does it take to assemble 2 computers?	**Rate Problem**—a problem in which an object travels at two rates or two objects travel at two different rates. The formula $d = rt$, where d is distance, r is rate, and t is time, is used to solve these problems. Example: A car travels 300 miles at an average rate of 64 mph. If the rate is 50 mph for the first 4 hours, what is the rate for the last 2 hours?

Use one of the key vocabulary words to complete each sentence.

1. Andrea can solve a ______________________ to find how long it takes two people to clean test tubes in the lab.

2. ______________________ use the formula $d = rt$.

3. To find how many mL of two solutions to mix to get a 500 mL solution that is 5% acetic acid, Molly needs to solve a ______________________.

CAHSEE Standards Mini-Review 1

Standards: 6PS1.0, 6PS2.0, 6PS3.0

You need to be able to

find the probability of independent events.

The names of 10 students are put in a hat. The student names include Linda and Harriet (one name each). One name is drawn, replaced, and then a second name is drawn. What are the chances of both names being Linda or her friend Harriet?

The two events are independent. The chances of Linda or Harriet being picked on either drawing are 2 out of 10. In simplest form, this is the fraction: ______

Multiply the two probabilities to find the chances of both names being Linda or Harriet. $\frac{1}{5} \cdot \frac{1}{5} =$ ______

What are the chances of both Linda and Harriet being picked?

Since 10 names can be picked on each drawing, the total number of outcomes is 10 • 10, or ______.

Two of these outcomes are Linda-Harriet and Harriet-Linda. The probability of both being picked is 2 out of 100. In simplest form, this is the fraction: ______

TEST PRACTICE 8

Bob has a number cube with the numbers 1–6 and two coins. He rolls the cube and tosses the coins. What is the probability of his getting a 3 and at least one head?

A $\frac{1}{12}$ **B** $\frac{1}{8}$ **C** $\frac{1}{6}$ **D** $\frac{1}{2}$

Hint: Find each of the probabilities first.

CAHSEE Standards Mini-Review 1

Statistics, Data Analysis, and Probability

You need to be able to

compute the mean, median, and mode of data sets.

Here are the ages of the people at a family picnic. Find the median, mode, and mean for this data set.

13	8	7	42	3	35	31	13	20
24	36	17	3	13	40	13	31	

Start by putting the data in order from least to greatest.

3 3 7 8 13 13 13 13 17
20 24 31 31 35 36 40 42

How many ages are in the data set? ______

The median is the middle age or the average of the two middle ages. Since 17 ÷ 2 = 8.5, the middle age when arranged in order is the 9th age in the list. What is this age? ______

The mode is the age which occurs most often. What is it? ______

Add the ages and divide by 17 to find the mean. The sum of the ages in this list is 349, so the mean is ______.

TEST PRACTICE 1

Barbara has quiz scores of 80, 85, 70, 75, and 80. Which of the following will give her a mean score of 80 on all six quizzes?

A 75 **C** 85
B 80 **D** 90

Hint: A mean score of 80 on 6 texts gives a total score of 480.

CAHSEE Standards Mini-Review 1

Standards: 6PS1.0, 6PS2.0, 6PS3.0

You need to be able to

understand the affects of outliers.

This list shows the salaries of 10 people in a small business.

Yearly Salaries at Eastlake Imports				
$22,000	**$22,000**	**$24,000**	**$26,000**	**$28,000**
$30,000	**$30,000**	**$45,000**	**$48,000**	**$185,000**

Which better describes this data, the median or the mean?

Find the median and the mean. Hint: The sum of the data is $460,000.

median = ______   mean = ______

The mean is too high to be a good representative for the data. This is because one salary is much higher than all the others. A data item such as this is called an *outlier*. It lies outside the other numbers.

Which better represents this data, the mean or the median? ______

TEST PRACTICE 2

Why is the mean NOT a good representative for this set of data?

Weights of Packages (oz)									
9	5	8	6	95	6	4	8	9	7

A It is too low.
B It is too high.
C It is less than the median.
D It is not a whole number.

Hint: When a data set has outliers, the median is a better representative than the mean.

CAHSEE Standards Mini-Review 1

Statistics, Data Analysis, and Probability

You need to be able to

find the probability of an event not happening.

Fred put these marbles in a bag. What are the chances of *not* getting a striped marble?

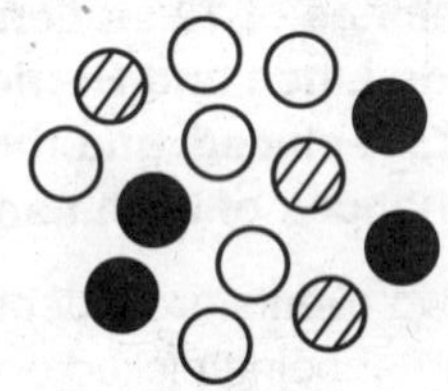

Write the number of white and black marbles over the total number of marbles. ______

You will either get a striped marble or not. The sum of the two probabilities must equal 1.

$P(\text{getting striped}) + P(\text{not getting striped}) = \frac{3}{13} + \frac{10}{13} = \frac{13}{13} = 1$

Julie found that $\frac{3}{8}$ of her neighbors have cell phones.

If a neighbor is chosen at random, what is the probability that the person does *not* have a cell phone?

Subtract $\frac{3}{8}$ from 1 to solve the problem. $1 - \frac{3}{8} = \frac{8}{8} - \frac{3}{8} =$ ______

TEST PRACTICE 7

The spinner shows the stuffed toys you can win in a game. If you spin the spinner, what is the probability of not winning a toy bear?

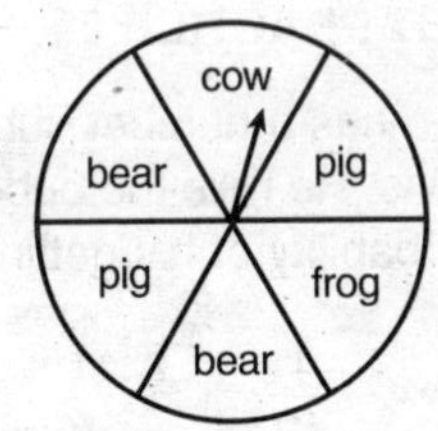

A $\frac{1}{3}$ **B** $\frac{1}{2}$ **C** $\frac{2}{3}$ **D** $\frac{5}{6}$

Hint: $P(\text{an event happening}) + P(\text{event not happening}) = 1$

CAHSEE Standards Mini-Review 1

Standards: 6PS1.0, 6PS2.0, 6PS3.0

> **You need to be able to**
>
> use probability to make predictions.

Tomas flipped a coin 25 times. He recorded his results in the table below. If Tomas flips the same coin a total of 100 times, how many times out of 100 do you predict the coin will land heads up?

Heads	Tails
卌 卌 |	卌 卌 ||||

Since Tomas has already conducted an experiment, you can use his data to make a more accurate prediction.

Tomas's coin landed heads up _____ times out of 25 flips.

Write a fraction to represent the number of times Tomas's coin has landed heads up so far. Then multiply by the total number of times Tomas will conduct his experiment.

_____ × 100 = _____

TEST PRACTICE 6

Margaret rolled a number cube 35 times and recorded her results in the table shown. How many more times, out of the next 60 rolls, do you predict Margaret will roll a number greater than 4?

A 10

B 20

C 24

D 32

Number	1	2	3	4	5	6
Times Rolled	卌	卌 |	||||	卌 |	卌 |||	卌 |

Hint: How many times out of 35 did Margaret roll a number greater than (but not equal to) 4?

CAHSEE Standards Mini-Review 1

Statistics, Data Analysis, and Probability

> **You need to be able to**
>
> explain why a sample might be biased.

Hillary asked each student in her class how many books he or she checked out of the city library over the summer. She plans to use the data to make a recommendation to the city council to close the library over the summer. Is her sample biased?

A *biased* sample is limited. It does not represent all points of view necessary to draw a valid conclusion.

Describe Hillary's sample.

____

Are there people Hillary should have included in her sample who might have a different point of view? If so, describe the people she should have included.

____

Is Hillary's sample biased? _____

TEST PRACTICE 3

Evon is taking a survey to see if there's interest in a city water park. Which of the following samples are NOT biased?

A students at Evon's school

B people coming out of a grocery store

C residents of the city senior center

D members of the swim team

Hint: When you consider each sample, think about points of view that may be missing.

CAHSEE Standards Mini-Review 1

Standards: 6PS1.0, 6PS2.0, 6PS3.0

You need to be able to

represent the outcomes for a probability experiment.

Miriam spins this spinner and tosses a penny and a dime. Each coin can land heads up (H) or tails up (T).

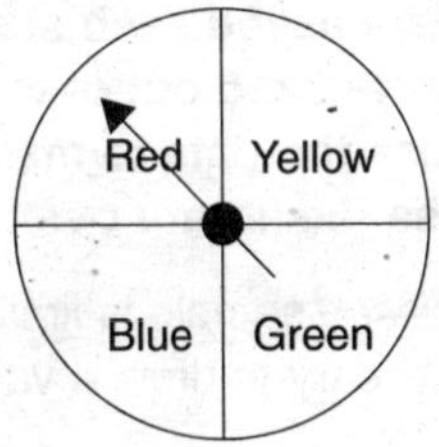

Make a tree diagram to show the outcomes. Use H for heads and T for tails.

Write the four colors in the rectangles. Then use H and T to fill in the squares. The diagram has been started for you.

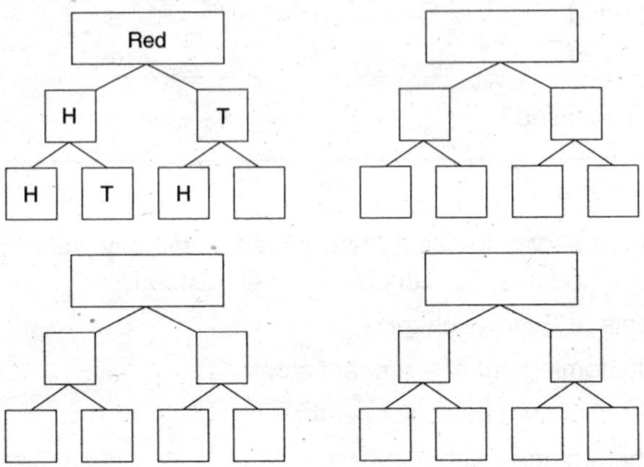

CAHSEE Standards Mini-Review 1

Statistics, Data Analysis, and Probability

Finish this list of all 16 outcomes for Miriam's experiment.

Red HH, Red HT, Red TH, ______________________

Yellow HH, ______________________

____

____

TEST PRACTICE 5

Josh does a probability experiment with two four-part spinners. One spinner has pictures of a duck, fox, cat, and dog; the other spinner has the colors red, blue, green, and yellow. Which tree diagram shows the sample space for this experiment?

A Duck — R; Fox — B; Cat — G; Dog — Y

B Duck — R R R; Fox — B B B; Cat — G G G; Dog — Y Y Y

C Duck — R B G Y; Fox — R B G Y; Cat — R B G Y; Dog — R B G Y

D Duck — R R R R; Fox — B B B B; Cat — G G G G; Dog — Y Y Y Y

Hint: To find the number of combinations shown by a tree diagram, count the items in the bottom row.

CAHSEE Standards Mini-Review 2

Standard: 7NS1.0

You need to be able to

Solve problems that involve percent.

A. Karen has finished 8 of the 15 math problems assigned for homework. What percent of her math homework has she finished?

Estimate first. Has she finished more or less than half? ________

Will the percent be more or less than 50%? ________

Find the percent. $\frac{8}{15} = \_\_\_\_\ \%$

B. Carlos has finished 73% of his reading assignment of 98 pages. Estimate how many pages he has read.

73% is about _____ or _____

98 is about _____ pages.

$\frac{3}{4} \cdot 100 = \_\_\_\_\_$

Carlos has read about _____ pages.

TEST PRACTICE 8

Which is the BEST estimate of 59% of 198?

A 12 **C** 120

B 100 **D** 1200

Hint: What percent is near 59% and easy to use?

CAHSEE Standards Mini-Review 2

Number Sense

You need to be able to

compare rational numbers.

Compare $-\frac{3}{8}$ and $-\frac{5}{8}$.

The denominators are the same, so compare the numerators.

$-3 \bigcirc -5$, so $-\frac{3}{8} \bigcirc -\frac{5}{8}$.

Order $-\frac{5}{8}, \frac{1}{4}, -\frac{7}{12}$, and $\frac{1}{3}$ from least to greatest.

Write the rational numbers with a common denominator of 24.

$-\frac{5}{8} = -\frac{5 \cdot \_\_\_}{8 \cdot \_\_\_} = -\frac{\_\_\_}{24}$ $\frac{1}{4} = \frac{1 \cdot \_\_\_}{4 \cdot \_\_\_} = \frac{\_\_\_}{24}$

$-\frac{7}{12} = -\frac{7 \cdot \_\_\_}{12 \cdot \_\_\_} = -\frac{\_\_\_}{24}$ $\frac{1}{3} = \frac{1 \cdot \_\_\_}{3 \cdot \_\_\_} = \frac{\_\_\_}{24}$

Order the numerators from least to greatest.

_____ < _____ < _____ < _____

From least to greatest, the rational numbers are

_____, _____, _____, _____.

TEST PRACTICE 1

Which numbers are in order from least to greatest?

A $\frac{2}{3}, \frac{5}{6}, -\frac{1}{2}, -\frac{7}{9}$ **C** $-\frac{1}{2}, -\frac{7}{9}, \frac{5}{6}, \frac{2}{3}$

B $-\frac{1}{2}, \frac{2}{3}, \frac{5}{6}, -\frac{7}{9}$ **D** $-\frac{7}{9}, -\frac{1}{2}, \frac{2}{3}, \frac{5}{6}$

Hint: What must you do before you can order the rational numbers by their numerators?

CAHSEE Standards Mini-Review 2

Standard: 7NS1.0

You need to be able to

write numbers in scientific notation using positive exponents.

Represent 6,850,000 in scientific notation.

Move the decimal point left until the value of the number is greater than or equal to 1 and less than 10.

6,850,000 =

6,850,000. =

Write the number of places you moved the decimal point as the power of 10.

$6.85 \cdot 10^{-}$

6,850,000 = ________ • ________

Write $2.7 \cdot 10^8$ in standard form.

Move the decimal point right the same number of places as the power of 10. Write zeros as needed.

2.700

$2.7 \cdot 10^8$ = ____________

TEST PRACTICE 2

Light travels about 186,000 miles per second. Which expression represents this number in scientific notation?

A $1.86 \cdot 10^3$

B $18.6 \cdot 10^4$

C $1.86 \cdot 10^5$

D $1.86 \cdot 10^6$

Hint: How many places left do you need to move the decimal point?

CAHSEE Standards Mini-Review 2

Number Sense

You need to be able to

convert fractions, decimals, and percents.

Write 2% as a decimal.

2% = .02. = 0.______

Write 0.9 as a percent.

0.9 = 0.90.% = ______%

Write 3 out of 8 as a percent.

3 out of 8 = $\frac{3}{8}$

3 ÷ 8 = _________, so $\frac{3}{8}$ = _________.

0.375 = 0.375% = ______.5%

TEST PRACTICE 7

The students in Mr. Field's math class described parts of the class in different ways. Which describes the greatest part of the class?

A 1 out of 8

B 0.18

C 8%

D 0.8

Hint: Which answer choices will you write in another form so that you can compare the choices?

CAHSEE Standards Mini-Review 2

Standard: 7NS1.0

You need to be able to

compute with integers.

What is the value of the sum –4 + 7?

Draw a number line and arrows to show the addition.

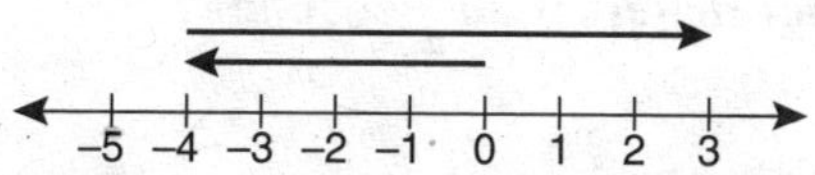

From the number line, –4 + 7 = ________.

What is the value of the difference –8 – (–1)?

Draw a picture of a jar with 8 negative signs in it.

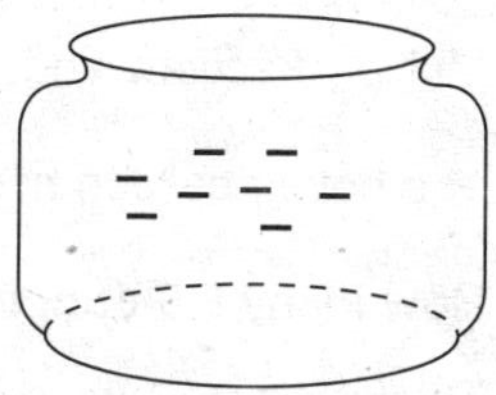

When one "–" is taken away, there are ________.

So, –8 – (–1) = ________.

What is the value of the quotient –8 ÷ 4?

When the 8 "–" in the jar above are divided into 4 equal groups, there are ________ "–" in each group. So, –8 ÷ 4 = ________.

TEST PRACTICE 6

Which of the following equals (–13)(–7)?

A –91 **C** –6

B –20 **D** 91

Hint: What is the sign of the product when both factors are negative numbers?

CAHSEE Standards Mini-Review 2

Number Sense

You need to be able to

write numbers in scientific notation using negative exponents.

Represent 0.000007 in scientific notation.

Move the decimal point right until the value of the number is greater than or equal to 1 and less than 10. 0.000007 → 0.000007.0

Write the number of places you moved the decimal point as the negative exponent in the power of 10. $7.0 \cdot 10^{-}$

0.000007 = ________ • ________

Write $1.15 \cdot 10^{-5}$ in standard form.

Move the decimal point left the same number of places as the absolute value of the power of 10. Write zeros as needed. .0001.15

$1.15 \cdot 10^{-5}$ = ________

TEST PRACTICE 3

Two units used to measure weights are pounds and tons. One pound equals $5.0 \cdot 10^{-4}$ ton. Which represents this number in standard form?

A 0.005 **C** 50,000

B 0.0005 **D** 0.00005

Hint: How many places left do you need to move the decimal point?

CAHSEE Standards Mini-Review 2

Standard: 7NS1.0

You need to be able to

add and subtract fractions and decimals.

A truck contains $1\frac{3}{5}$ tons of sand. It receives $1\frac{1}{2}$ tons more. How many tons are in the truck altogether?

Add $1\frac{3}{5}$ and $1\frac{1}{2}$ because ______________________

The GCF of 5 and 2 is ______. $1\frac{3}{5} = 1\frac{}{10}$ $1\frac{1}{2} = 1\frac{}{10}$

So, $1\frac{3}{5} + 1\frac{1}{2} =$ ________. The truck has ______ tons of sand.

Wanda needs 4 pounds of fish. The scale shows 3.675 pounds. How much more fish is needed?

Subtract 3.675 from 4 because ____________________.

Perform the subtraction. $4.000 - 3.675 =$ ________

Put ________ more pounds of fish on the scale.

TEST PRACTICE 4

Kyla is $54\frac{3}{5}$ inches tall and Diana is $53\frac{1}{4}$ inches tall. How many inches taller is Kyla than Diana?

A $1\frac{7}{20}$ in. **B** $1\frac{2}{5}$ in. **C** $1\frac{3}{5}$ in. **D** $1\frac{4}{9}$ in.

Hint: What operation should you use?

CAHSEE Standards Mini-Review 2

Number Sense

You need to be able to

multiply and divide fractions and decimals.

A hot air balloon slowly rises $3\frac{1}{2}$ feet per second. How high up is the balloon after rising for $30\frac{3}{4}$ seconds?

Multiply $3\frac{1}{2}$ by $30\frac{3}{4}$ because ______________________

Change mixed numbers to fractions. $3\frac{1}{2} \cdot 30\frac{3}{4} =$ ____ • ____

$=$ ____ $=$ ____

Multiply fractions and write the result as a mixed number.

The hot air balloon rises ________ feet.

How many 0.5-ounce containers are needed to hold 20.5 ounces?

Think: Two 0.5 ounce containers will hold ________.

Divide the total number of ounces by 0.5. $0.5\overline{)20.5}$

The number of containers needed is ______.

TEST PRACTICE 5

A letter weighs 0.85 ounces. How many ounces will a stack of 150 of these letters in a mailing weigh?

A 12.75 ounces **C** 127.5 ounces

B 124.5 ounces **D** 1,275 ounces

Hint: Estimate to help you rule out incorrect answers.

CAHSEE Standards Mini-Review 3

Standard: 7NS2.0

You need to be able to

understand absolute value.

What is the value of $2|3 \cdot (-6)|$?

Multiply inside the absolute value signs.

$2\,|3 \cdot (-6)| = 2\,|$ ______ $|$

Take the absolute value of -18.

$2\,|-18| = 2 \cdot$ ______

Multiply by 2.

$2 \cdot 18 =$ ______

Use >, <, or = to compare. **$-3\,|6 - 10|$ ______ $-2\,|-4 - 3|$**

Simplify each expression. Then compare.

$-3\,|6 - 10| = -3 \cdot 4 =$ ______ $-2\,|-4 - 3| = -2 \cdot 7 =$ ______

$-12 > -14$, so $-3\,|6 - 10|$ ______ $-2\,|-4 - 3|$

TEST PRACTICE 8

Which expression has the greatest value?

A $|12 - 15|$ **B** $|-8 - 5|$ **C** $3\,|-6 + 2|$ **D** $-3\,|8 - 9|$

Hint: The absolute value is always a positive number. Choice D cannot be correct because it is the only negative choice.

CAHSEE Standards Mini-Review 3

Number Sense

You need to be able to

understand positive exponents.

Evaluate the expression $4^3 + 6^2$.

In the expression 4^3, 4 is the base and 3 is the exponent. An exponent shows how many times the base is used as a factor.

Write each expression as a product. Multiply, then add.

$4^3 + 6^2 = (4 \times 4 \times 4) +$ ______ $=$ ______ $+ 36 =$ ______

Find the value of $\left(\frac{2}{5}\right)^3 \div \frac{4}{25}$.

Evaluate $\left(\frac{2}{5}\right)^3$. $\left(\frac{2}{5}\right)^3 = \frac{2 \times 2 \times 2}{5 \times 5 \times 5} =$ ______

Use the reciprocal of the divisor. Then multiply.

$\left(\frac{2}{5}\right)^3 \div \frac{4}{25} = \frac{8}{125} \div \frac{4}{25} = \frac{8}{125} \times \frac{25}{4} =$ ______

TEST PRACTICE 1

What is the value of $\left(\frac{5}{6}\right)^2 \times \left(\frac{3}{10}\right)^3$?

A $\frac{1}{16}$ **B** $\frac{3}{16}$ **C** $\frac{45}{4}$ **D** $\frac{3}{160}$

Hint: Remember to use the exponent for both terms in a fraction.

$\left(\frac{a}{b}\right)^2 = \frac{a^2}{b^2}$ $\left(\frac{a}{b}\right)^n = \frac{a^n}{b^n}$

CAHSEE Standards Mini-Review 3

Standard: 7NS2.0

You need to be able to

understand negative exponents.

Evaluate the expression $4^{-3} \times 2^5$.

Rewrite 4^{-3} as a fraction with a positive exponent. $4^{-3} = \frac{1}{4^3}$

Evaluate each exponential expression. Then multiply.

$4^{-3} \times 2^5 = \frac{1}{4^3} \times 2^5 = \frac{1}{64} \times 25 =$ _____

Find the value of $\frac{7}{8} + \left(\frac{2}{3}\right)^{-3}$.

Evaluate $\left(\frac{2}{3}\right)^{-3}$. $\left(\frac{2}{3}\right)^{-3} = \left(\frac{3}{2}\right)^{3}$ _____

Add the numerators. Reduce to simplest form.

$\frac{7}{8} + \left(\frac{2}{3}\right)^{-3} = \frac{7}{8} + \left(\frac{3}{2}\right)^{3} = \frac{7}{8} + \frac{27}{8} =$ _____ $=$ _____

TEST PRACTICE 2

What is the value of $\left(\frac{5}{8}\right)^{-2}$?

A $\frac{10}{16}$ **B** $\frac{25}{8}$ **C** $\frac{25}{64}$ **D** $\frac{64}{25}$

Hint: A negative exponent shows a reciprocal.

$a^{-2} = \frac{1}{a^2}$ $\left(\frac{a}{b}\right)^{-n} = \left(\frac{b}{a}\right)^{n}$

CAHSEE Standards Mini-Review 3

Number Sense

You need to be able to

estimate a square root.

Which point on the number line best represents $\sqrt{7}$?

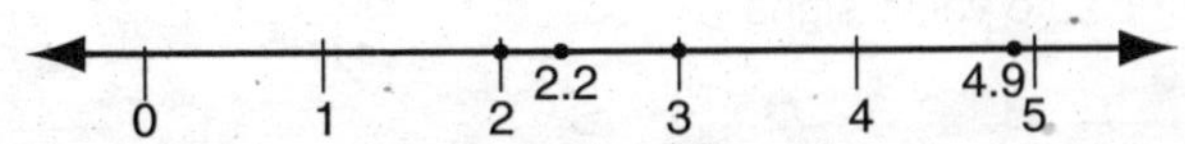

Perfect squares close to 7 are 4 and _____.

$\sqrt{4} =$ _____ and $\sqrt{9} =$ _____, so $\sqrt{7}$ is between _____ and _____.

The point for _____ on the number line is between 2 and 3.

What is a good estimate of $\sqrt{85}$?

Perfect squares close to 85 are _____ and _____.

$\sqrt{81} =$ _____ and $\sqrt{100} =$ _____.

85 is closer to 81 than to 100, so _____ is a good estimate of $\sqrt{85}$.

TEST PRACTICE 7

A square has an area of 52 square meters. What is the best estimate of the length of each side of this square?

A 7 meters

B 8 meters

C 13 meters

D 26 meters

Hint: How is the side of a square related to its area?

CAHSEE Standards Mini-Review 3

Standard: 7NS2.0

You need to be able to
find the square root of a perfect square integer.

What is the value of 25^2?

The area of a square whose sides are 25 units long is 25^2.

25

25

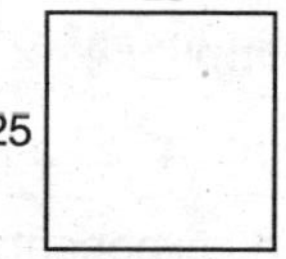

So, $25^2 =$ ______ • ______ = ______

What is the square root of 169?

Look for the dimensions of a square whose area is 169 square units. Complete each equation below with "=" or "≠"

length × width = area

7 • 7 ____ 169   9 • 9 ____ 169   12 • 12 ____ 169   13 • 13 ____ 169

A square whose length and width are ____ has an area of 169.

$\sqrt{169}$ is ______ .

TEST PRACTICE 6

Which statement is true?

A $83^2 = 2 \cdot 83$

B $\sqrt{225} = 15$

C $\sqrt{84} = 42$

D $\sqrt{4} = 16$

Hint: Remember the difference between the square of a number and the square root of a number.

CAHSEE Standards Mini-Review 3

Number Sense

You need to be able to
add fractions using common denominators.

Add: $\frac{7}{8} + \frac{5}{6}$.

Rewrite the fractions with a common denominator.
Use the least common multiple (LCM) of 8 and 6.

multiples of 8: ______, ______, ______, ______, ______

multiples of 6: ______, ______, ______, ______, ______

What is the LCM of 8 and 6? ______

$\frac{7}{8} \to \frac{7\,\bullet\,\_\_\_\_}{8\,\bullet\,\_\_\_\_} \to \frac{\_\_\_\_}{24}$ $\frac{5}{6} \to \frac{5\,\bullet\,\_\_\_\_}{6\,\bullet\,\_\_\_\_} \to \_\_\_\_$

Add the numerators. Write the common denominator.
Then simplify the sum.

$\frac{21}{24} + \frac{20}{24} = \frac{\_\_\_}{} = \frac{24}{24} + \frac{\_\_\_}{24} = 1\frac{\_\_\_}{24}$

TEST PRACTICE 3

In science class, Lou spent $\frac{3}{4}$ hour performing an experiment. He spent another $\frac{1}{3}$ hour doing research. How long did he spend on science?

A $\frac{15}{12}$h   **B** $\frac{4}{7}$h   **C** $1\frac{1}{12}$h   **D** $1\frac{1}{4}$h

Hint: Do you need to find a common denominator?

CAHSEE Standards Mini-Review 3

Standard: 7NS2.0

You need to be able to

subtract fractions using common denominators.

Subtract: $4\frac{1}{5} - 1\frac{3}{4}$.

Rewrite the fractions with a common denominator.

$4\frac{1}{5} \longrightarrow 4\frac{}{20} \longrightarrow 3\frac{}{20}$

$1\frac{3}{4} \longrightarrow 1\,\underline{} \longrightarrow 1\,\underline{}$

Rename so you can subtract twentieths.

Think: $4\frac{4}{20} = 3 + 1 + \frac{4}{20}$

$= 3 + \frac{}{20} + \frac{4}{20}$

$3\frac{24}{20} - 1\frac{15}{20} =$ ______

TEST PRACTICE 4

Two snow storms brought heavy snow to one ski area. The first brought $6\frac{1}{2}$ inches of snow. The second snowstorm brought $9\frac{3}{8}$ inches of snow. How much more snow fell during the second snowstorm?

A $2\frac{7}{8}$ in. **B** $3\frac{1}{8}$ in. **C** $3\frac{1}{3}$ in. **D** $3\frac{7}{8}$ in.

Hints: Do you need to do any renaming?

CAHSEE Standards Mini-Review 3

Number Sense

You need to be able to

use exponent rules to simplify expressions.

Simplify $4^2 \times 3 \times 2^3 \times 3^5$.

Write 4^2 as a power of 2. $4^2 = 4 \times 4 = 2 \times 2 \times 2 \times 2 = 2^4$

Add exponents for factors with the same base.

$2^4 \times 3 \times 2^3 \times 3^5 = (2^4 \times 2^3) \times (3 \times 3^5) = 2^7 \times$ ______

Simplify $\frac{2^3 \cdot 3^4 \cdot 5^3}{2 \cdot 3^6 \cdot 5^2}$.

Write the expression as a product of factors with the same base. $\frac{2^3 \cdot 3^4 \cdot 5^3}{2 \cdot 3^6 \cdot 5^2} = \frac{2^3}{2} \cdot \frac{3^4}{3^6} \cdot \frac{5^3}{5^2} =$

Simplify each factor, then multiply. $\frac{2^2}{1} \cdot \frac{1}{3^2} \cdot \frac{5}{1} = \frac{4 \cdot 5}{9} =$ ______

TEST PRACTICE 5

$\frac{6^2 \cdot 3^4 \cdot 2^2}{3^2 \cdot 6^3} =$

A 4 **B** 6 **C** 8 **D** 216

Hint: Remember these rules for products and quotients of powers.

Product of Powers	Quotient of Powers
$a^n \times a^m = a^{n+m}$	$\frac{a^n}{a^m} = a^{n-m} = \frac{1}{a^{n-m}}$

CAHSEE Standards Mini-Review 4

Standards: 7AF1.0, 7AF2.0

You need to be able to

interpret the meaning of graphs.

These graphs show the amount of money in four savings accounts.

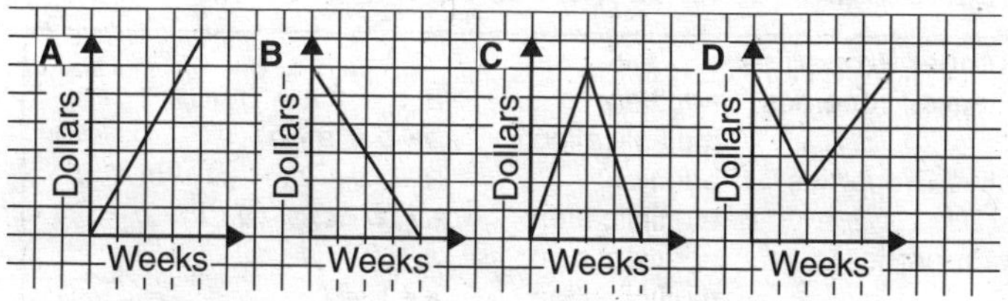

Write the letter of the graph that matches each situation.

Pauline deposited money for a while, then made some withdrawals. _____

Michael took out money for a couple of weeks, then began saving again. _____

Edna is saving money steadily. _____

Wallace is making the same withdrawal every week. _____

TEST PRACTICE 8

The graph shows the money in a savings account. Choose the description that best matches the graph.

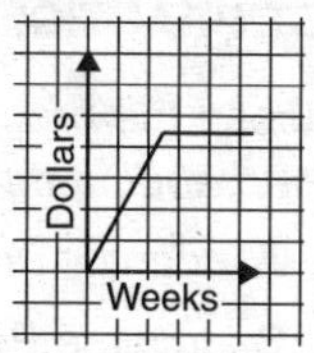

A Mae saved steadily, then made withdrawals.

B Mae saved steadily, then stopped.

C Mae made withdrawals, then stopped.

D Mae made withdrawals, then begin to save.

Hint: What does a horizontal line on the line graph indicate?

CAHSEE Standards Mini-Review 4

Algebra and Functions

You need to be able to

use letters as variables in mathematical expressions.

Tom had 924 stamps in his collection, including duplicates. He gave an equal number of duplicate stamps to 2 of his friends. Write a math expression that shows the stamps Tom has left.

Describe the situation in words.

→ number of stamps Tom had minus number of friends times number of ____________

→ Substitute ________ for number of stamps Tom had.

→ Substitute 2 for ____________.

→ Substitute s for ____________.

stamps Tom had	minus	number of friends	times	stamps Tom gave each friend
↓		↓		↓
____	◯	____	◯	____

Without a multiplication sign, you can write this expression as ____________________.

TEST PRACTICE 1

Catherine bought 3 packages of ball point pens and 2 fountain pens. Each package contains the same number of pens. Which best represents the number of pens Catherine bought?

A $3 + 2p$ **B** $3p + 2$ **C** $3p + 2p$ **D** $3(p + 2)$

Hints: How could you find the total number of pens?

CAHSEE Standards Mini-Review 4

Standards: 7AF1.0, 7AF2.0

You need to be able to

write an equation for a verbal description.

CANOE RENTALS

Basic fee	$5.00
Hourly rate	$4.50

The sign shows the cost of renting a canoe. Write an equation that can be used to find *c*, the cost in dollars, of renting a canoe for 1 hour.

Note: You do not have to include dollar signs in the equation because *c* represents numbers of dollars.

Cost equals ________ plus rate for ________ hours.

Write an equation that can be used to find *c*, the cost in dollars, of renting a canoe for:

3 hours ________

h hours ________

TEST PRACTICE 2

Which equation can be used to find *c*, the cost in dollars of renting a campsite for *d* days?

CAMPSITE RENTAL

Basic fee	$8.00
Daily rate	$6.50

A $c = 8 + 6.5$

B $c = 8d + 6.5d$

C $c = 8 + 6.5d$

D $c = 8d + 6.5$

Hint: What stays the same no matter the number of days?

CAHSEE Standards Mini-Review 4

Algebra and Functions

You need to be able to

name the property that justifies a process used.

These properties are true for all real numbers.

Real-Number Properties		
Commutative Properties	addition:	$a + b = a + b$
	multiplication:	$a \cdot b = a \cdot b$
Associative Properties	addition:	$a + (b + c) = (a + b) + c$
	multiplication:	$a \cdot (b \cdot c) = (a \cdot b) \cdot c$
Identity Properties	addition:	$a + 0 = a$
	multiplication:	$a \cdot 1 = a$
Inverse Properties	addition:	$a + -a = 0$
	multiplication:	$a \cdot \frac{1}{a} = 1$
Distributive Property		$a \cdot (b + c) = (a \cdot b) + (a \cdot c)$

Write the name of the property that justifies each step.

$5 \cdot \frac{3}{4} \cdot \frac{1}{5} = \frac{3}{4} \cdot 5 \cdot \frac{1}{5}$ ________

$(\frac{3}{4} \cdot 5) \cdot \frac{1}{5} = \frac{3}{4} \cdot (5 \cdot \frac{1}{5})$ ________

$\frac{3}{4} \cdot (5 \cdot \frac{1}{5}) = \frac{3}{4} \cdot 1$ ________

$\frac{3}{4} \cdot 1 = \frac{3}{4}$ ________

TEST PRACTICE 7

Choose the property for addition that matches this description.
When adding three numbers, you can add any two of them first.

A Identity Property

B Inverse Property

C Commutative Property

D Associative Property

Hint: Write three addends and show that you can add any two first.

CAHSEE Standards Mini-Review 4

Standards: 7AF1.0, 7AF2.0

You need to be able to

name the property illustrated by an equation.

Here are some properties that are true for all real numbers.

Commutative Properties	addition:	$a + b = a + b$
	multiplication:	$a \cdot b = a \cdot b$
Associative Properties	addition:	$a + (b + c) = (a + b) + c$
	multiplication:	$a \cdot (b \cdot c) = (a \cdot b) \cdot c$
Identity	addition:	$a + 0 = a$
	multiplication:	$a \cdot 1 = a$
Distributive		$a \cdot (b + c) = (a \cdot b) + (a \cdot c)$

Complete this equation to show the Distributive Property.

$(37 \cdot 5) + (23 \cdot 5) = (\_\_\_\_\_\_) \cdot 5$

Write the name of the property that matches each description.

You can multiply two numbers in either order. ______

When adding three numbers,
you can add any two of them first. ______

TEST PRACTICE 6

Choose the property illustrated by this equation.

$(2.1 \cdot 0.8) \cdot 7.4 = 7.4 \cdot (2.1 \cdot 0.8)$

A Commutative **C** Distributive

B Associative **D** Identity

Hint: Which property deals with the order of factors?

CAHSEE Standards Mini-Review 4

Algebra and Functions

You need to be able to

write an expression or an equation for a verbal description.

Larry and his 3 friends share a restaurant bill equally. Write an expression to show each person's share. Use *B* for the total.

This situation involves division.
Show that *B* is divided into 4 equal parts. ______

Let *d* stand for the distance Fred hiked. Write an expression to show how far he'll have gone after another 15 miles.

This is an addition situation. Show that 15 is added to *d*. ______

Show that Katy (*k*) is 4 years younger than Ralph (*r*).

The relationship between their ages uses
subtraction. Show that *k* is 4 less than *r*. $k =$ ______

Write an equation to show that Hal (*h*) sold three times as many magazines as his friend Alice (*a*).

The relationship between their sales uses
multiplication. Show that *h* is equal to 3 times *a*. $h =$ ______

TEST PRACTICE 3

Edwardo is 46 years old. Which equation shows that he is 27 years older than his daughter Tammy?

A $E + 27 = T$ **C** $E = 27 - T$

B $E = T + 27$ **D** $E = T - 27$

Hint: What will you need to add to Tammy's age to find her father's age?

CAHSEE Standards Mini-Review 4

Standards: 7AF1.0, 7AF2.0

You need to be able to

use the order of operations to simplify expressions.

Simplify $10 + 20 \div 4 \cdot 3 - 2$.

	$10 + 20 \div 4 \cdot 3 - 2$
Multiply and divide from left to right.	$10 +$ ______ $\cdot 3 - 2$
Add and subtract from left to right.	$10 +$ ______ $- 2$
	______ $- 2$
	______

Simplify $16 \div (4 - 2) + 3$.

	$16 \div (4 - 2) + 3$
Perform the operation inside parentheses.	$16 \div$ ______ $+ 3$
Multiply and divide from left to right.	______ $+ 3$
Add and subtract from left to right.	______

TEST PRACTICE 4

Simplify $15 + 10 \div (5 - 3) \cdot 4$.

A 8 **B** 20 **C** 35 **D** 80

Hints: When will you do the operation inside the parentheses? In what order do you multiply, divide, add, and subtract?

CAHSEE Standards Mini-Review 4

Algebra and Functions

You need to be able to

use the order of operations to evaluate expressions.

Find the value of $4a^2 - 10(a - 1)$ for $a = -3$.

Substitute -3 for the variable *a*.	$4(-3)^2 - 10(-3 - 1)$
Perform the operation inside parentheses.	$4(-3)^2 - 10($______$)$
Raise -3 to the second power.	$4 \cdot$ ______ $- 10(-4)$
Simplify.	$36 - (-40) =$ ______

Find the value of $xy^2 + 2(x + 10)$ for $x = 5$ and $y = -2$.

Substitute 5 for the variable *x*.	$5y^2 + 2(5 + 10)$
Simplify the second term.	$5y^2 +$ ______
Substitute -2 for the variable y.	$5($______$)^2 + 30$
Simplify.	$5 \cdot 4 + 30 =$ ______

TEST PRACTICE 5

What is the value of $3p(10 - 2q)$ when $p = -1$ and $q = 3$?

A -72 **B** -39 **C** -12 **D** 39

Hints: Watch out for parentheses! The expressions $5y^2$ and $(5y)^2$ do not have the same value.

CAHSEE Standards Mini-Review 5

Standards: 7AF3.0, 7AF4.0

You need to be able to

solve problems involving direct variation.

Use the table to write an expression to find the capacity of a container in quarts that has a capacity of *n* gallons.

Container Capacity

Gal	2	3	4	5	n
Qt	8	12	16	20	?

What can you do to each number of gallons to get the number of quarts? ______________

Describe the number of quarts in terms of the number of gallons. ______________

Write the expression. ______________

TEST PRACTICE 8

Which expression can be used to find the length of a room in yards that has a length of *n* feet?

Length (in feet)	6	9	12	15	n
Length (in yards)	2	3	4	5	?

A $3n$ **B** $4n$ **C** $\frac{n}{3}$ **D** $\frac{n}{4}$

Hint: How do you change feet to yards?

CAHSEE Standards Mini-Review 5

Algebra and Functions

You need to be able to

graph linear functions.

The table gives the relationship between the weight of a chicken and the time it takes to cook. Use the table to make a graph.

Chicken Weight (lb)	Cooking Time (h)
5	2
6	$2\frac{1}{4}$
7	$2\frac{1}{2}$
8	$2\frac{3}{4}$
9	3

Complete the graph to show this.

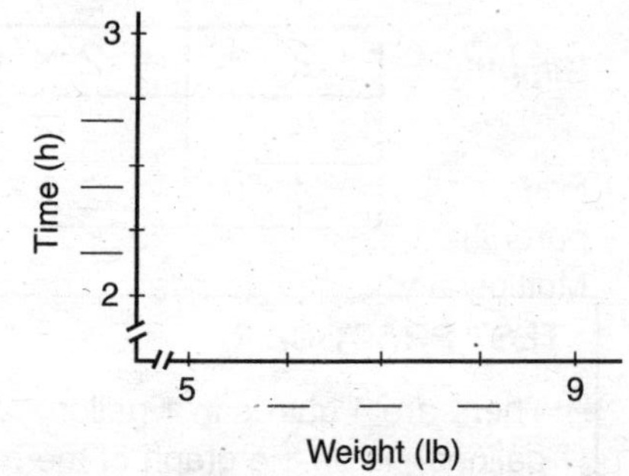

What point indicates the time it takes to cook an 8-pound chicken?

____

Do you think that you could extend the graph to 20 pounds?

____

TEST PRACTICE 1

Joan is 3 times her sister's age plus 2 years. Which expression shows Joan's age?

A $3x$ **C** $3x + 2$

B $2x + 3$ **D** $3x - 2$

Hint: Is Joan younger or older than her sister?

CAHSEE Standards Mini-Review 5

Standards: 7AF3.0, 7AF4.0

You need to be able to

graph linear functions.

George invented the unit "gorg". Each gorg is equivalent to 2 feet. Sketch a graph to represent the number of gorgs *g* in *f* feet.

Complete the table to show how many gorgs *g* are in *f* feet. Then graph the ordered pairs. Draw a line from the origin through the points.

feet	gorgs
1	$2 \times 1 = 2$
2	$2 \times$ ___ = ___
3	$2 \times$ ___ = ___
4	$2 \times$ ___ = ___

TEST PRACTICE 2

There are 4 quarts in 1 gallon. Which ordered pair, (quarts, gallons), is on the graph of the relationship that gives the number of gallons in a given number of quarts?

A (64, 16) **C** (64, 32)

B (8, 64) **D** (32, 64)

Hint: Can you test ordered pairs? For example, how would you find the number of gallons in 64 quarts? Is it $64 \div 4$, $64 \cdot 4$, $64 \div 2$, or some other quotient or product?

CAHSEE Standards Mini-Review 5

Algebra and Functions

You need to be able to

solve problems using speed and rate.

Mr. Miller drove 435.2 miles at a constant speed. He drove for a total of 8 hours. At what speed in miles per hour did he drive?

Complete the division problem at the right using the given information. $\frac{\text{miles}}{\text{hours}}$

Divide the numerator by the denominator. $8\overline{)435.20}$

Write the speed in miles per hour. ______ miles per ______

Janice can buy 6 cans of juice for $7.50. What is the unit price, or cost per can of juice?

Complete this division using the words "cost" and "number of cans." ____) ____

Divide using numbers for the words.

The unit price is $______ per can of juice.

TEST PRACTICE 7

Two dozen (24) notebooks cost $32.00. What is the cost of one notebook?

A $0.75 per notebook **C** $1.33 per notebook

B $0.85 per notebook **D** $16.00 per notebook

Hint: To calculate a unit rate, find the cost per notebook.

CAHSEE Standards Mini-Review 5

Standards: 7AF3.0, 7AF4.0

You need to be able to

solve and graph inequalities.

Graph $m < 2$ on a number line.

Show all numbers less than 2 on a number line.
Use an open dot to show that 2 is not included in the solution set.

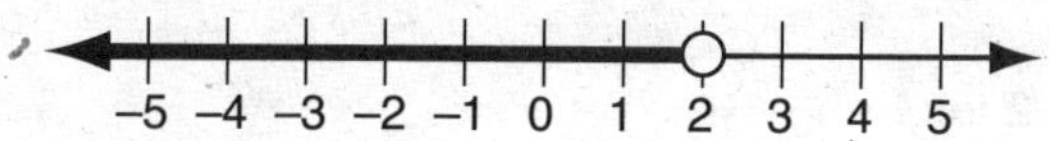

Solve $2y \geq -6$. Graph the solution set on a number line.

Divide both sides by 2 to isolate the variable *y*.

$2y \div 2 \quad \geq -6 \div 2$

$y \geq$ ______

Show all numbers greater than or equal to -3 on a number line.
Use a closed dot to show that -3 is included in the solution set.

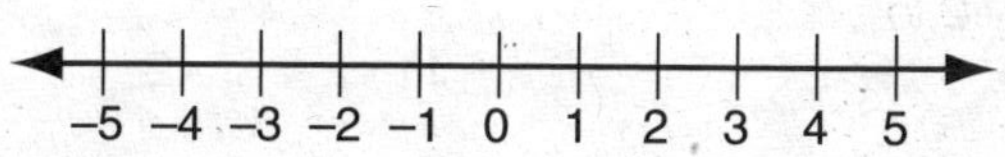

TEST PRACTICE 6

What is the solution set for $1 - 2a > 7$?

A $a < -16$ **C** $a < -4$ **B** $a < -12$ **D** $a < -3$

Hint: If you multiply both sides of an inequality by a negative number, you must reverse the inequality symbol. The inequalities $-2a > 6$ and $2a < -6$ are equivalent.

CAHSEE Standards Mini-Review 5

Algebra and Functions

You need to be able to

graph linear functions.

A car travels at the speed shown in the table. Write an equation that best represents the relationship in the table.

Time (h)	1	2	3	4	___
Distance (mi)	62	124	186	___	310

The distance equals the time times the rate. What is the rate or speed? ______

How far would the car travel in 8 hours at the same rate? ______

$D = r \cdot t$. Substitute the rate for *r*. $D =$ ______ $\cdot\, t$

Make a graph that shows the information in the table.

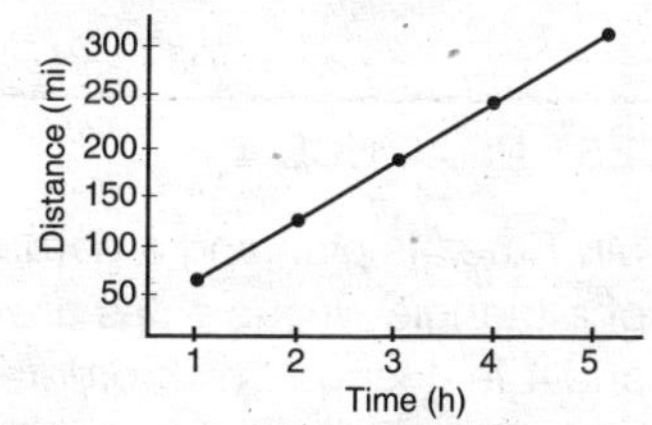

What will your ordered pairs be?

______

TEST PRACTICE 3

Which of the following points does NOT belong on the graph of $y = 4x + 2$?

A (0, 2) **C** (0.5, 4)

B (2, 10) **D** (3, 15)

Hint: If $x = 0$, what is *y*?

CAHSEE Standards Mini-Review 5

Standards: 7AF3.0, 7AF4.0

You need to be able to

graph linear functions.

Sasha is planning to make a rectangular banner 3 feet wide. She has not decided how long to make it. Graph an equation that gives perimeter *P* in terms of length *x*.

Sketch a rectangle that has length x and width 3.

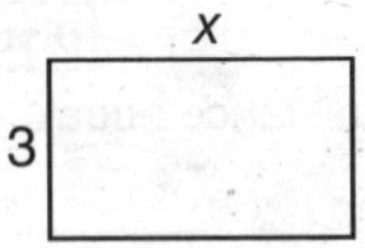

Write an equation for perimeter, P. $P = 2(x + 3)$

Make a table of values. Use the table to sketch the graph.

x	y
1	8
2	10
3	12

TEST PRACTICE 4

Mr. Timez is planning a garden in the shape of a triangle whose sides are x feet, 4 feet, and 4 feet long. Which ordered pair would be on the graph of the equation that gives perimeter P in terms of x?

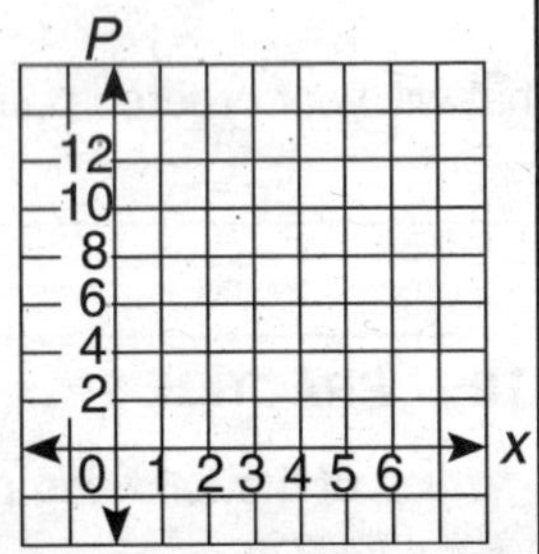

A (3, 10) **C** (3, 11)

B (3, 4) **D** (3, 22)

Hint: To graph a relationship, you will need an equation, a table of values, or other data you can use to plot points. What is the equation you should use for Test Practice 4?

CAHSEE Standards Mini-Review 5

Algebra and Functions

You need to be able to

solve two-step equations.

Solve for n if $2n + 3 = 17$.

Subtract 3 from both sides to isolate the variable term. $2n + 3 - 3 = 17 - 3$

$2n =$ ______

Divide both sides by 2 to isolate the variable n. $2n \div 2 = 14 \div 2$

$n =$ ______

Solve $8 = \frac{a}{2} - 3$.

Add 3 to both sides to isolate the variable term. $8 + 3 = \frac{a}{2} - 3 + 3$

______ $= \frac{a}{2}$

Multiply both sides by 2 to isolate the variable a. $11 \cdot 2 = \frac{a}{2} \cdot 2$

______ $= a$

TEST PRACTICE 5

What is the solution to the equation $10 - 3x = 16$?

A -18 **B** $-8\frac{2}{3}$ **C** -2 **D** $2\frac{2}{7}$

Hint: Check your solution by substituting it back into the original equation. Simplify. The expressions on the left and right sides of the equal sign must be equal.

CAHSEE Standards Mini-Review 6

Standards: 7MG1.0, 7MG2.0, 7MG3.0

TEST PRACTICE 7

Which net can be used to make this three-dimensional shape?

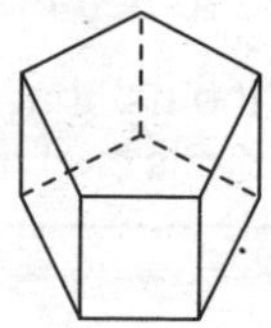

A

C

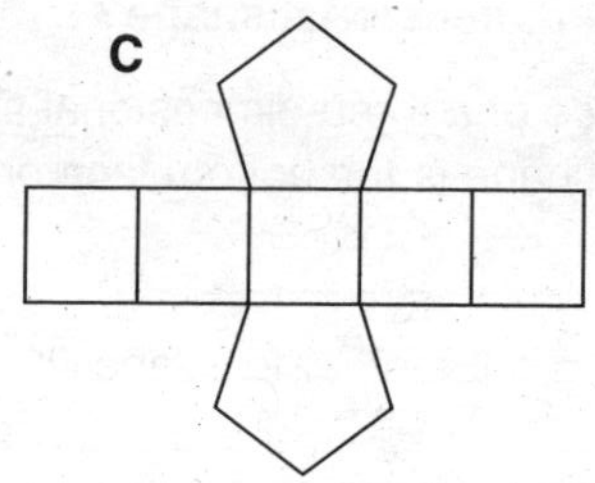

B

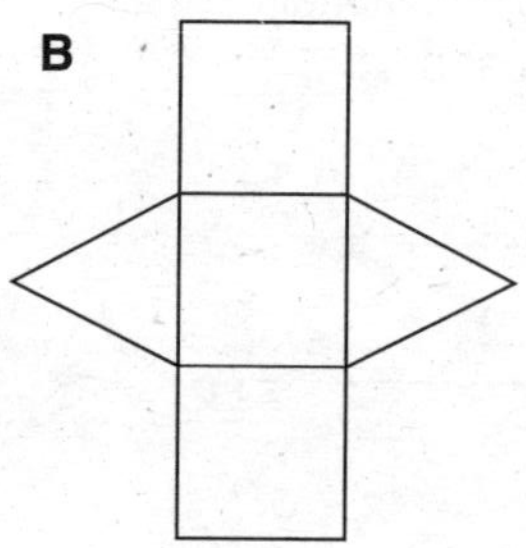

D

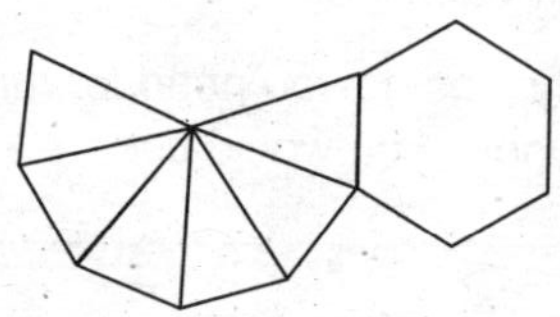

Hint: To make a net for a three-dimensional shape, start by identifying the polygons that are used for its faces.

CAHSEE Standards Mini-Review 6

Measurement and Geometry

You need to be able to

compute perimeter or circumference.

Art walks twice around the jogging path shown. How far does he walk?

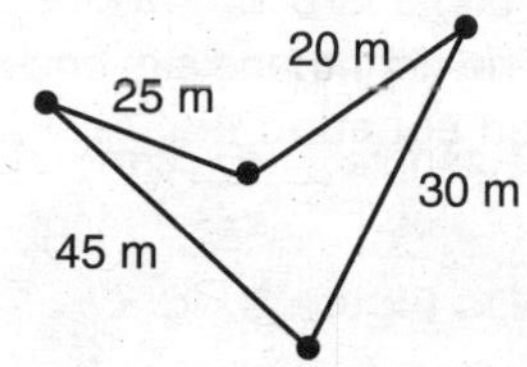

The problem asks for the perimeter of the polygon. Add the four side lengths.
25 m + 20 m + 30 m + 45 m = ______ m

Multiply by 2 because Art walks around the perimeter twice.
2 • 120 m = ______ m

About how much fence is needed for a circular garden with a diameter of 16 ft?

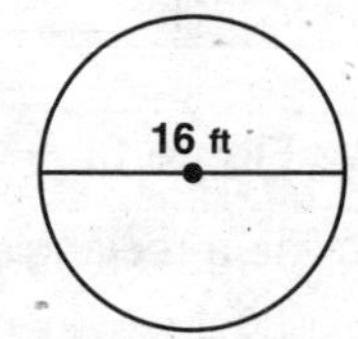

The circumference of a circle equals the diameter times pi. Pi (π) is approximately equal to 3.14.
You can use 3 for pi to make estimates.

3 • 16 ft = ______ ft

TEST PRACTICE 1

How much trim is needed to go around the edges of a rectangular rug that measures 5 ft wide and 8 ft long?

A 13 ft **C** 39 ft
B 26 ft **D** 40 ft

Hint: To find the perimeter of a rectangle, add the length to the width. Then multiply by 2.

CAHSEE Standards Mini-Review 6

Standards: 7MG1.0, 7MG2.0, 7MG3.0

You need to be able to

use formulas to solve problems.

Use Figure A. Each square is 1 square centimeter.

The right triangle is one-half of a 6-by-8 rectangle. Find each area.

rectangle _____ cm^2 right triangle _____ cm^2

Use Figure B.

Draw a vertical line to divide the parallelogram into right triangles. Each triangle is one-half of a 3-by-8 rectangle.

one triangle _____ cm^2 parallelogram _____ cm^2

Use Figure C.

Outline a rectangle 8 cm tall and 6 cm wide.

The area of the triangle is one-half of this rectangle.

rectangle _____ cm^2 triangle _____ cm^2

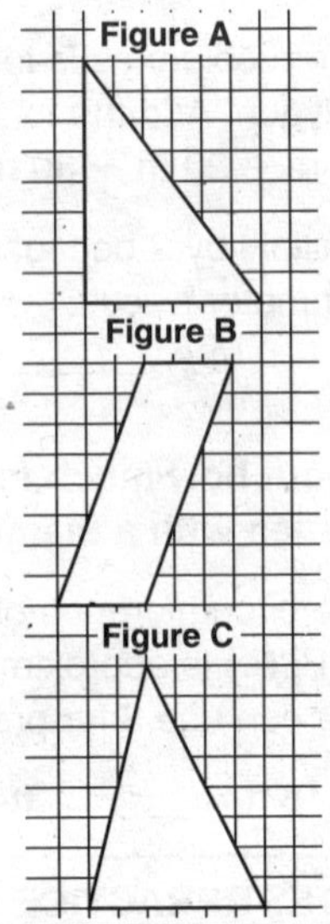

TEST PRACTICE 2

Which of these triangles has the same area as a 12-inch square? The base and height of each triangle are given.

A 4 in., 36 in. **C** 8 in., 9 in.

B 8 in., 18 in. **D** 16 in., 18 in.

Hint: The area of a triangle is half the product of its base and height.

CAHSEE Standards Mini-Review 6

Measurement and Geometry

You need to be able to

identify the net that can be used to make a three-dimensional model.

What polygons do you need to make a net for this three-dimensional shape?

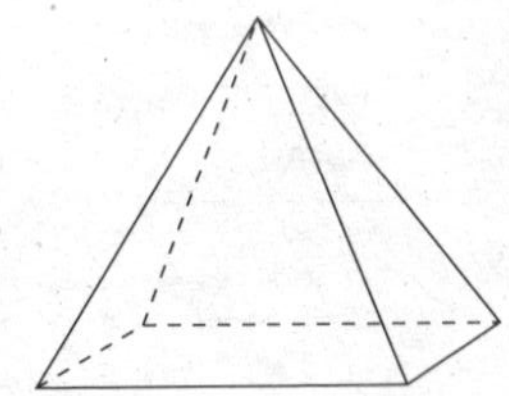

The flat surface of a three-dimensional shape is a *face*. What shape is the bottom face or base?

____

What shapes are the other four faces?

____

A *net* is a pattern you can use to make a three-dimensional shape. The net for a three-dimensional shape must have one polygon for each face.

How many polygons do you need for this net? _____ Name the polygons you would use.

____

What three-dimensional figure can you make with each of these nets?

Net A Net B Net C

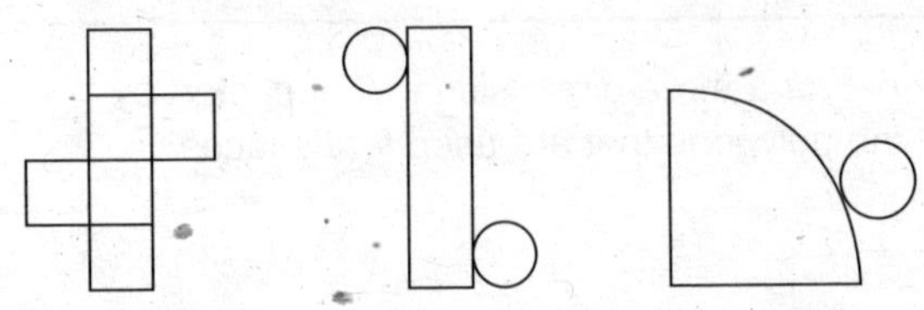

_______ _______ _______

CAHSEE Standards Mini-Review 6

Standards: 7MG1.0, 7MG2.0, 7MG3.0

You need to be able to

solve problems using the Pythagorean Theorem.

Clare drew a pattern for a quilt piece. The pattern must be a right triangle. What can Clare do to make sure her pattern is a right triangle?

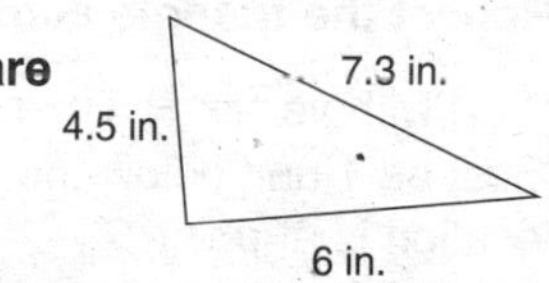

Write the formula for the Pythagorean Theorem.

____

Find the angle that looks like a right angle.
Label the sides of the pattern *a*, *b*, and *c*.

Substitute the measures of sides *a*, *b*, and *c* into the Pythagorean Theorem and simplify both sides.

$\_\_\_^2 + \_\_\_^2 = \_\_\_^2$

$\_\_\_ = \_\_\_$

If the values on both sides of the equal sign are equal, the triangle is a right triangle. Is Clare's pattern a right triangle?_____

TEST PRACTICE 6

How long must the diagonal of this frame be in order for the corners to be right angles?

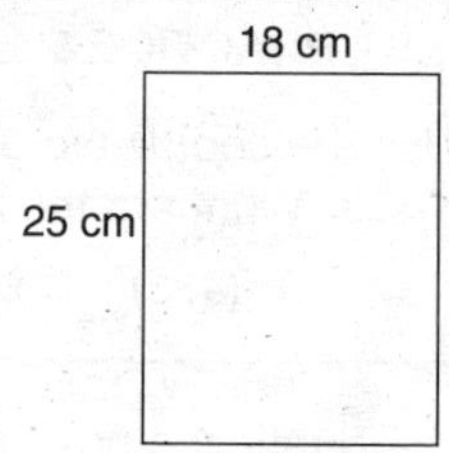

A 43 cm **C** 17.3 cm

B 30.8 cm **D** 19.3 cm

Hint: Draw a diagonal of the frame. Look for a right triangle. Is the diagonal a leg or hypotenuse of the triangle?

CAHSEE Standards Mini-Review 6

Measurement and Geometry

You need to be able to

use formulas to solve problems.

Match each formula to its description.

s = side, *l* = length, *w* = width, *h* = height, *B* = area of the base

Volume (cube)	$6s^2$
Volume (rectangular prism)	$2lw + 2lh + 2wh$
Surface area (cube)	lwh
Surface area (rectangular prism)	Bh
Volume (any prism)	s^3

Hints: Any formula with *s* as a variable refers to a cube.

The surface area of a figure is calculated in square units.

A rectangular prism can have up to 3 unique rectangular faces. These faces come in pairs. Finding the area of a face and multiplying it by 2 gives you the area of 2 of the 6 faces.

Since *B* represents the area of the base, any formula with the variable *B* is a general formula.

TEST PRACTICE 3

Which formula finds the volume of the figure shown?

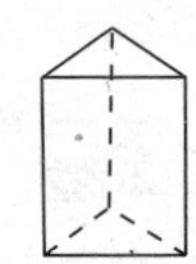

A $\frac{1}{2}Bh$ **C** lwh

B Bh **D** s^3

Hint: *B* represents the area of the base.

CAHSEE Standards Mini-Review 6

Standards: 7MG1.0, 7MG2.0, 7MG3.0

You need to be able to

solve problems involving scale factors and area.

Mrs. Johnson owns farmland that measures 500 m by 375 m. She is dividing the property into 4 equal-sized regions for family members by making dividing lines from the midpoint of each side of the property.

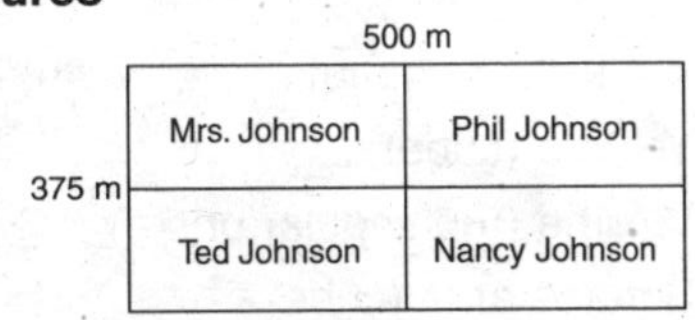

What will be the dimensions of Mrs. Johnson's property after dividing the property into 4 regions? ________

What is the scale factor for the reduction of Mrs. Johnson's property? ________

What can you multiply the original area by to find the area of Mrs. Johnson's smaller property? ________

TEST PRACTICE 4

A new school will have a length and width $\frac{8}{7}$ times its original measurements. Which will be the area of the school after construction?

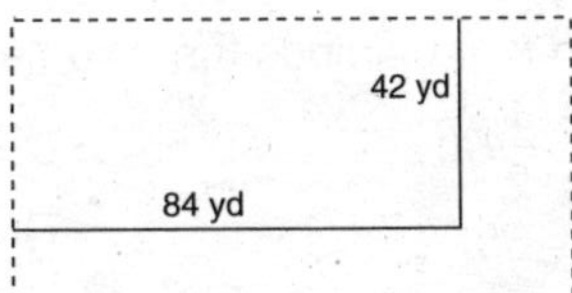

A 3087 yd^2
B 3528 yd^2
C 4032 yd^2
D 4608 yd^2

Hint: Remember that the scale factor affects both length and width.

CAHSEE Standards Mini-Review 6

Measurement and Geometry

You need to be able to

use a coordinate graph to reflect a figure.

Reflect the triangle across the *x*-axis.

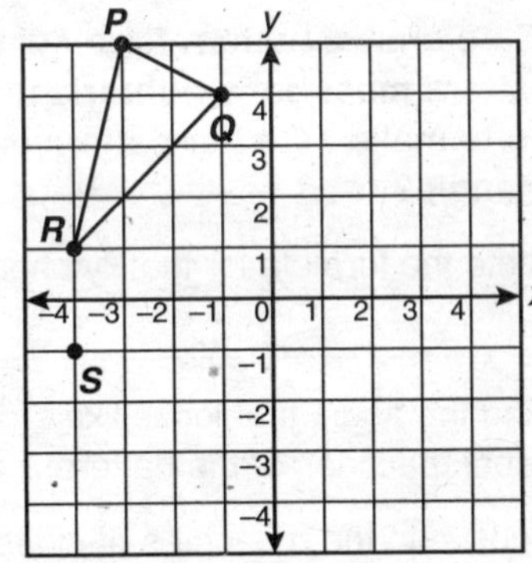

Start with vertex *R*. The reflected point must be 1 unit below the *x*-axis. The new location is at point *S*.

Reflect point *Q* across the *x*-axis. The reflection should be 4 units below the *x*-axis. What are the new coordinates? ________

Reflect point *P* across the *x*-axis. The reflection should be 5 units below the *x*-axis. What are the new coordinates? ________

Connect the points to make the reflected triangle.

If you reflect point *S* across the *y*-axis, what happens to the *x*-coordinate? ________________

TEST PRACTICE 5

Use the coordinate plane above. Reflect $\triangle PQR$ across the *y*-axis. What are the new coordinates of point *R*?

A (4, 1) **B** (4, −1) **C** (−4, −1) **D** (1, 4)

Hint: To reflect a point across the *y*-axis, change the sign of the *x*-coordinate.

CAHSEE Standards Mini-Review 7

Standard: 7PS1.0

You need to be able to

use a scatterplot to make predictions.

Alex recorded the number of inches of rain and the growth of a sunflower each week for 10 weeks. He displayed his results on the scatterplot shown. Make a prediction. How many inches do you think the sunflower will grow during a week with 6 inches of rain?

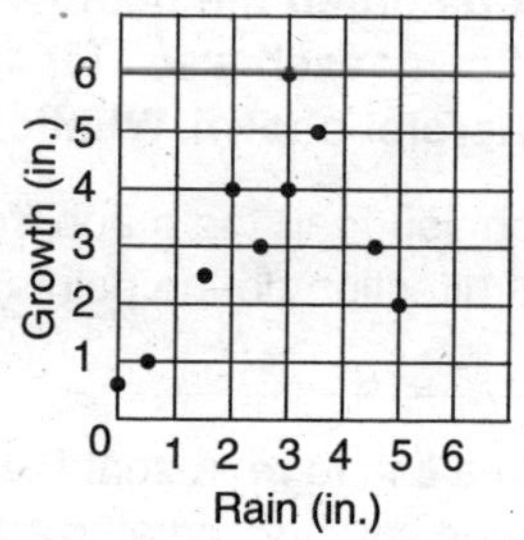

Follow the points on the graph to make a prediction. The points seem to be dropping after 3 inches of rainfall in one week. Six inches of rainfall will likely produce

_____ growth than 5 inches of rainfall.

Prediction: _____ inches of growth.

TEST PRACTICE 8

Look at Alex's scatterplot. Which is the best prediction for the amount of rainfall during a week in which the sunflower grew 7 inches?

A 2 in.

B 3 in.

C 5 in.

D 6 in.

Hint: How many inches of rain produced the greatest amount of growth in one week?

CAHSEE Standards Mini-Review 7

Statistics, Data Analysis, and Probability

You need to be able to

find the minimum, maximum, and median of a data set.

Here are the ages of the people at a family picnic.

13	8	7	42	3	35	31	15	20
24	36	17	3	10	40	13	31	

What is the median age?

Start by putting the data in order from least to greatest.

3 3 7 8 10 13 13 15 17
20 24 31 31 35 36 40 42

How many ages are there in the data set? _____

Since $17 \div 2 = 8.5$, the middle age when arranged in order is the 9th age in the list. This age is 17. There are 8 people younger than 17, and 8 people older than 17.

TEST PRACTICE 1

Use the family picnic ages. Add one more person who is 45 years old. What is the new median age?

A 17 **C** 18.5

B 17.5 **D** 20

Hint: If there is an even number of data items, the median is the average of the two middle numbers.

CAHSEE Standards Mini-Review 7

Standard: 7PS1.0

You need to be able to

find the upper and lower quartiles of a data set.

Here are the final test scores in Ms. Hansen's history class.

70 60 60 80 90 65 85 75 55 90 65 65 75 95 90

What is the lower quartile score?

Put the data in order from least to greatest. Circle the median score.

55 60 60 65 65 65 70 75 75 80 85 90 90 90 95

The lower quartile score is the median of the lower half of the data.

55 60 60 (65) 65 65 70 (75) 75 80 85 90 90 90 95

What is the upper quartile score?

The upper quartile score is the median of the upper half of the data. Circle the upper quartile score.

55 60 60 (65) 65 65 70 (75) 75 80 85 90 90 90 95

TEST PRACTICE 2

Use this set of test scores. What is the lower quartile score?

55 60 60 65 65 75 80 85 90 90 90

A 60 **B** 62.5 **C** 75 **D** 90

Hint: The median and the two quartiles divide a data set into 4 equal parts.

CAHSEE Standards Mini-Review 7

Statistics, Data Analysis, and Probability

You need to be able to

describe data represented on a scatterplot.

Alex recorded the number of inches of rain and the growth of a sunflower each week for 10 weeks. He displayed his results on the scatterplot shown. What effect did the rain have on the sunflower?

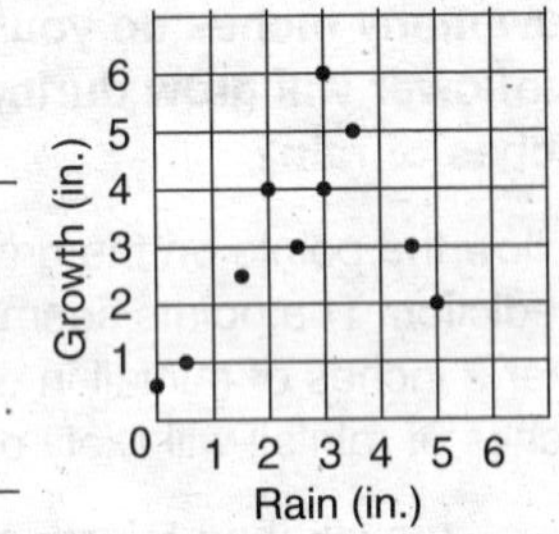

When you read the graph from left to right, in what direction do the points move?

____

Draw a line to represent the "average" points. Is it vertical, horizontal, diagonal with top to the right or diagonal with top to the left?

____

Horizontal lines indicate no effect. Diagonal lines going up from left to right indicate a positive effect. Diagonal lines going down from left to right indicate a negative effect. What effect did rain have on the growth of Alex's sunflower?

____

TEST PRACTICE 7

Look at Alex's scatterplot. Which conclusion can you make?

A Morê rain is always better.

B Sometimes too much rain limits growth.

C Less rain is always better.

D Growth is not effected by rain.

Hint: Look at the points for weeks with more than 3 inches of rain.

CAHSEE Standards Mini-Review 7

Standard: 7PS1.0

You need to be able to

choose the best type of graph to display a set of data.

Here are four different types of graphs.

Double Line Graph

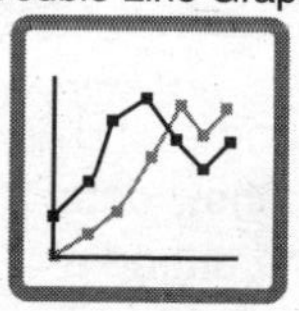

Double Bar Graph

Circle Graph

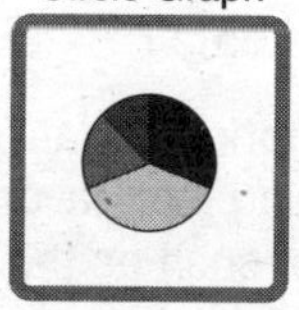

Line Plot

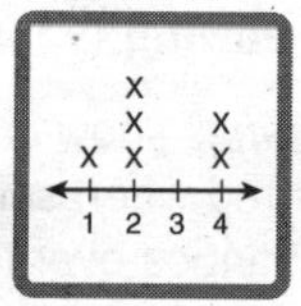

Line graphs show changes over time. Bar graphs illustrate comparisons. A circle graph shows a whole divided into parts. A line plot shows frequency of data on a number line.

A class votes for one of two choices for the class play. Which graph type could compare the results for the girls with the results for the boys? ________

The best graph for showing the ages of every high school student on a cheerleading squad is ________.

The best graph for showing the changes in populations in two cities is ________.

TEST PRACTICE 6

Which type of graph would be BEST to use to show the different amounts spent in a student's budget?

A circle **C** double line

B line plot **D** double bar

Hint: The student's budget is a whole divided into parts.

CAHSEE Standards Mini-Review 7

Statistics, Data Analysis, and Probability

You need to be able to

use a circle graph or a histogram to display data.

Fred recorded the age of people who went to see one showing of a movie. He displayed his data in both a circle graph and a histogram.

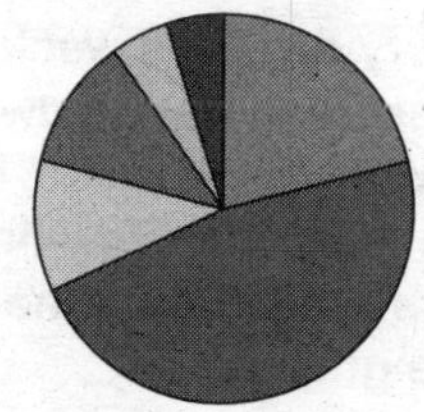

Were more or less than half the people at the showing ages 20-29?

____

On which graph is that fact easiest to see?

____

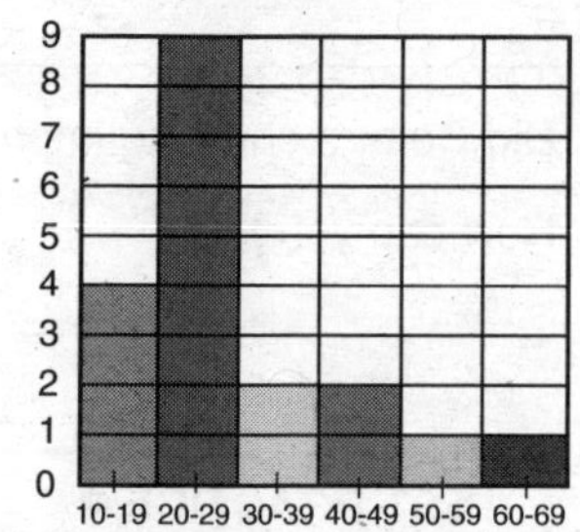

Tickets are $7. How much money did the showing earn? ________

Which graph allows you to calculate the amount the showing earned? ________

TEST PRACTICE 3

Which data would NOT be appropriate on a histogram?

A a student's test scores

B number of people at Summerfest at different times

C ages of the siblings of each student in your class

D average monthly temperatures each month in a city

Hint: Look for a set of data that could not be organized by ranges.

CAHSEE Standards Mini-Review 7

Standard: 7PS1.0

You need to be able to

use a line plot or a line graph to display data.

Every day for a week, Colby shot 10 basketball free throws and recorded how many he made. He wants to display the data to show whether his shooting improved. Should he use a line plot or a line graph?

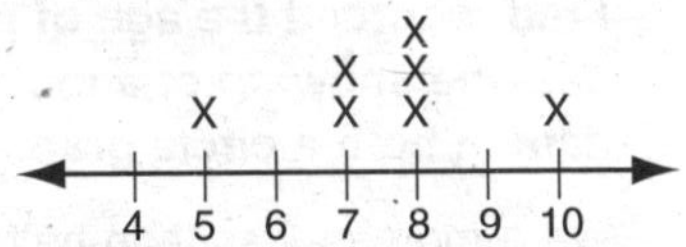

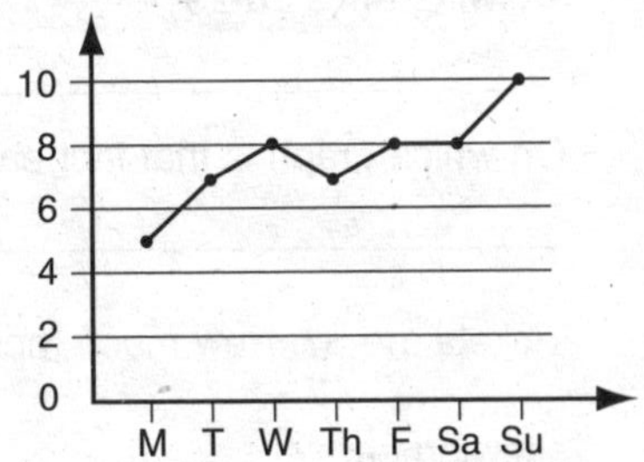

Which graph shows change over time?

____

Did Colby's shooting improve? ________

How can you tell?

____

TEST PRACTICE 4

Consider the line plot above. Which of the following measures can you immediately see?

A mean

B mode

C median

D quartiles

Hint: Which measure can you determine from the line plot without calculation or counting?

CAHSEE Standards Mini-Review 7

Statistics, Data Analysis, and Probability

You need to be able to

choose the best type of graph to display a set of data.

Which type of graph is most appropriate to show each of the following types of data?

Bar graphs show categories of data, like the number of people who chose red when selecting their favorite color. *Line graphs* show data that changes over time, such as the number of people who attend a show each day.

Speeds recorded of 25 cars traveling on a highway	bar graph
Number of responses to four choices of a survey question	circle graph
Every type of expense on which a week's paycheck is spent	line graph
Average daily temperatures recorded for one week	line plot

TEST PRACTICE 5

During a science experiment, Vashon recorded the temperature of a chemical mixture every 10 minutes. Which graph is most appropriate to show the data collected?

A bar graph

B circle graph

C line graph

D line plot

Hint: What is the key clue that will help you determine which graph to use?

CAHSEE Standards Mini-Review 8

Standards 1A1.0–1A5.0, 1A24.0, 1A25.0

You need to be able to

solve an inequality involving absolute value.

Solve $|g| < 4$. Graph the solution set on a number line.

Write two equivalent inequalities. $g < 4$ and $g > -4$

Show all numbers less than 4 *and* greater than -4.

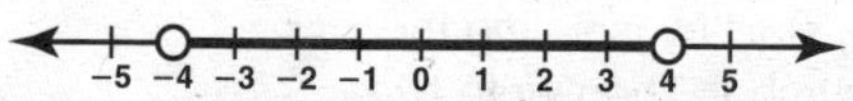

Solve $|y| > 2$. Graph the solution set on a number line.

Write two equivalent inequalities. $y > 2$ or $y < -2$

Show all numbers greater than 2 *or* less than -2.

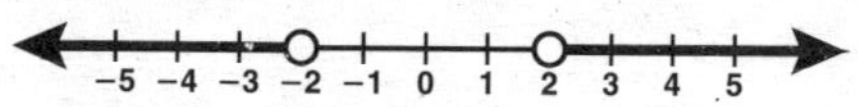

TEST PRACTICE 8

What is the solution set of the inequality $|7 - 2b| > 3$?

A $b > -5$ and $b < 2$ **C** $b < -2$ or $b > 5$

B $b > 2$ and $b < 5$ **D** $b < 2$ or $b > 5$

Hint: $|7 - 2b| > 3$ is equivalent to these two inequalities.
$7 - 2b > 3$ or $7 - 2b < -3$

CAHSEE Standards Mini-Review 8

Number Properties, Operations, and Linear Equations

You need to be able to

identify examples of the real number properties.

These properties are true for all real numbers.

Real-Number Properties		
Commutative Properties	addition:	$a + b = a + b$
	multiplication:	$a \cdot b = a \cdot b$
Associative Properties	addition:	$a + (b + c) = (a + b) + c$
	multiplication:	$a \cdot (b \cdot c) = (a \cdot b) \cdot c$
Identity Properties	addition:	$a + 0 = a$
	multiplication:	$a \cdot 1 = a$
Inverse Properties	addition:	$a + (-a) = 0$
	multiplication:	$a \cdot \frac{1}{a} = 1 \quad a \neq 0$
Distributive Property	$a \cdot (b + c) = (a \cdot b) + (a \cdot c)$	

Identify the property of real numbers shown by this equation.

$\mathbf{3(a + b^2) = 3a + 3b^2}$ The equation shows that a product equals a sum. The factor 3 is multiplied by each of the addends a and b^2.

This is an illustration of the ________________ Property.

TEST PRACTICE 1

Which multiplication property is illustrated by this equation?

$$\left(\frac{1}{2} + \frac{2}{3}\right) \cdot \left(\frac{3}{5} + \frac{1}{10}\right) = \left(\frac{3}{5} + \frac{1}{10}\right) \cdot \left(\frac{1}{2} + \frac{2}{3}\right)$$

A Commutative **C** Identity

B Associative **D** Inverse

Hint: To identify the Commutative Property, look for two addends or factors that switch places from the left to the right side.

CAHSEE Standards Mini-Review 8

Standards 1A1.0–1A5.0, 1A24.0, 1A25.0

You need to be able to

Identify the hypothesis and conclusion in a statement.

Identify the hypothesis and conclusion in a statement.

If a person lives in Los Angeles, then the person lives in California.

Is the statement in the form "If p, then q?" ________

The part of the statement following the word *If* is the hypothesis.
Write the hypothesis: ______________________

The part of the statement following the word *then* is the conclusion.
Write the conclusion: ______________________

TEST PRACTICE 2

What is the conclusion of this statement?
A plant is a flower if the plant is a daisy or a rose.

- **A** The plant is a flower.
- **B** The plant is a daisy.
- **C** The plant is a rose.
- **D** The plant is a daisy or rose

Hint: Rewrite the statement putting the *If* part first.

CAHSEE Standards Mini-Review 8

Number Properties, Operations, and Linear Equations

You need to be able to

solve an equation involving absolute value.

An absolute value equation is equivalent to a disjunction, two equations joined by the word *or.* For example, $|w| = 5$ is equivalent to $w = 5$ or $w = -5$. The solution set for $|w| = 5$ is $\{5, -5\}$.

To solve an absolute value equation, start by isolating the expression with the absolute value symbol. Then write two equivalent equations.

Solve $|2b| - 5 = -2$.

Add 5 to both sides to isolate the absolute value expression. $|2b| - 5 + 5 = -2 + 5$

$|2b| =$ ______

Write two equivalent equations. Solve each one for b.

$2b = 3$	$2b = -3$
$2b \div 2 = 3 \div$ ______	$2b \div 2 = -3 \div 2$
$b = \frac{3}{2}$	$b =$ ______

Use set brackets { } to write the two solutions. __________

TEST PRACTICE 7

What is the solution set of the equation $|2m + 1| = 5$?

A $\{2, -3\}$ **B** $\{2, -2\}$ **C** $\{3, -3\}$ **D** $\{8, -12\}$

Hint: $|2m + 1| = 5$ is equivalent to these two equations.
$2m + 1 = 5$ or $2m + 1 = -5$

CAHSEE Standards Mini-Review 8

Standards 1A1.0–1A5.0, 1A24.0, 1A25.0

You need to be able to

simplify and then solve a linear inequality in one variable.

Solve $8y \geq 6(y - 1)$. Graph the solution set on a number line.

Multiply by 6 to remove parentheses on the right side.	$8y \geq 6y - 6$
Subtract 6*y* from both sides.	$8y - 6y \geq 6y - 6 - 6y$
	$2y \geq$ ______
Divide both sides by 2.	$2y \div 2 \geq -6 \div 2$
	$y \geq$ ______

Show all numbers greater than or equal to −3 on a number line. Use a closed dot to show that −3 is included in the solution set. (An open dot would show that −3 is not included.)

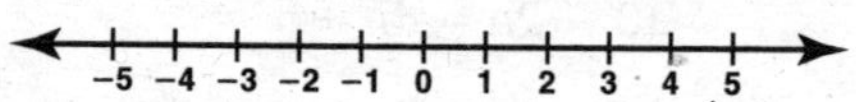

TEST PRACTICE 6

For what values of *m* is the following inequality true?

$-10 < 2(4 - 3m)$

A $m < -3$ **B** $m < 3$ **C** $m > -3$ **D** $m > 3$

Hint: If you multiply or divide by a negative number, you must reverse the inequality symbol.

CAHSEE Standards Mini-Review 8

Number Properties, Operations, and Linear Equations

You need to be able to

write the opposite or the reciprocal of a number.

Find the reciprocal of 0.4.

Remember, the product of a real number and its reciprocal is 1.

Change the decimal to a fraction. Reduce to simplest form.

$0.4 = \frac{4}{10}$ $\frac{4}{10} = \frac{4 \div 2}{10 \div 2} =$ ______

Interchange the numerator and the denominator.

The reciprocal of $\frac{2}{5}$ is ______.

Find the opposite of $p - 1$.

Multiply the entire expression times −1. Use the Distributive Property. Remember, the sum of a real number and its opposite is 0.

$-1 \cdot (p - 1) = -1 \cdot (p + (-1))$

$= (-1 \cdot p) + (-1 \cdot -1)$

$= -p +$ ______

TEST PRACTICE 3

How can you find the reciprocal of a number *n*?

A Multiply *n* by 1. **C** Divide *n* by 1.

B Multiply *n* by −1. **D** Divide 1 by *n*.

Hint: The product of reciprocals is 1.

CAHSEE Standards Mini-Review 8

Standards 1A1.0–1A5.0, 1A24.0, 1A25.0

You need to be able to

understand and use the rules of exponents.

These rules for exponents can be used to simplify expressions.

Product of Powers	Quotient of Powers	Power of a Power
$a^n \cdot a^m = a^{n+m}$	$\frac{a^n}{a^m} = a^{n-m} = \frac{1}{a^{m-n}}$	$(a^m)^n = a^{n\,m}$

Evaluate $2^4 \cdot 3^2 \cdot 2^{-6}$.

Add exponents for the base 2. $2^4 \cdot 3^2 \cdot 2^{-6} = 2^{-2} \cdot 3^2$

Write 2^{-2} as a fraction. $2^{-2} \cdot 3^2 = \frac{1}{2^2} \cdot 3^2$

Square the terms. $\frac{1}{2^2} \cdot 3^2 =$ ______

Simplify $\frac{6x^4}{8x}$.

Divide the coefficients by 2. $\frac{6x^4}{8x} = \frac{6x^4 \div 2}{8x \div 2} =$ ______

Subtract the exponents of the x-terms. $\frac{3x^4}{4x} = \frac{3}{4}x^{4-1} =$ ______

TEST PRACTICE 4

Which expression is equivalent to $a^3 \cdot a^4$?

A $a^2 \cdot a^6$ **B** $a \cdot a^6$ **C** $a \cdot a^7$ **D** $a^7 \cdot a^{12}$

Hint: To multiply expressions with the same base, add the exponents.

CAHSEE Standards Mini-Review 8

Number Properties, Operations, and Linear Equations

You need to be able to

simplify and then solve a linear equation in one variable.

Simplify both sides of an equation as much as possible before you solve it. Use the order of operations when you simplify.

1	2	3	4
Perform computations inside parentheses.	Evaluate exponential expressions.	Multiply and divide from left to right.	Add and subtract from left to right.

Solve for n in the equation $3(6 + 2n) - 8n = 9 + 4(n - 1)$.

Remove parentheses by using the Distributive Property. $18 + 6n - 8n = 9 + 4n - 4$

Combine terms on each side. $18 - 2n = 4n +$ ______

Add $2n$ to both sides. $18 - 2n + 2n = 4n + 5 + 2n$

Subtract 5 from both sides. $18 - 5 = 5 + 6n -$ ______

Divide sides by 6. $13 = 6n$

______ $= n$

TEST PRACTICE 5

Solve for x in the equation $3x - (6 + 5x) = 2(5 + x)$.

A $\frac{16}{7}$ **B** $-\frac{16}{3}$ **C** $\frac{8}{3}$ **D** -4

Hint: A negative sign in front of an expression shows multiplication by -1. For example $-(2x - 3) = -2x + 3$.

CAHSEE Standards Mini-Review 9

Standards 1A6.0–1A9.0

You need to be able to

graph the solution set for a system of two inequalities.

Solve this system of inequalities by graphing. $\begin{cases} y < x + 2 \\ x + 2y + 2 \geq 0 \end{cases}$

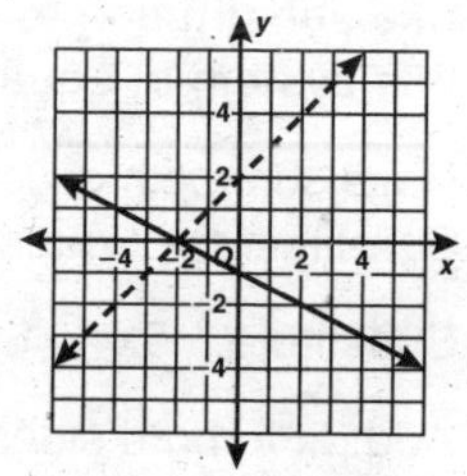

Graph a dotted line for the equation $y = x + 2$.

Graph a solid line for the equation $x + 2y + 2 = 0$. First convert it to slope-intercept form.

Graph $y =$ ________

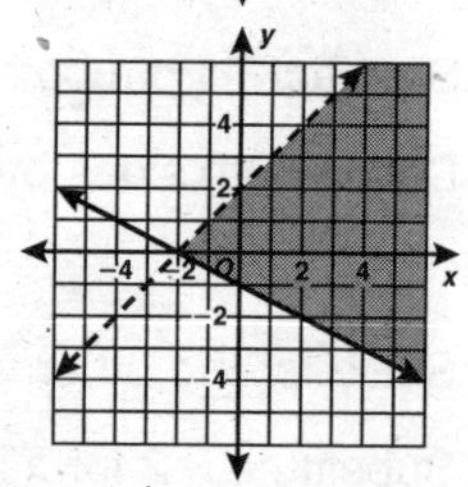

The two lines divide the plane into four regions.

Shade the region that shows the solution set for *both* inequalities.

TEST PRACTICE 8

Which system of inequalities is shown by the shaded region in the figure?

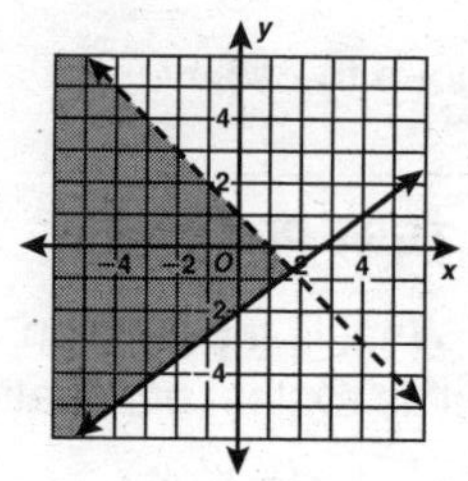

A $\begin{cases} y < 1 - x \\ 2x - 3y \geq 6 \end{cases}$ **C** $\begin{cases} y > 1 - x \\ 2x - 3y \geq 6 \end{cases}$

B $\begin{cases} y < 1 - x \\ 2x - 3y \leq 6 \end{cases}$ **D** $\begin{cases} y > 1 - x \\ 2x - 3y \leq 6 \end{cases}$

Hint: Which system includes (0, 0) in its solution?

CAHSEE Standards Mini-Review 9

Graphing and Systems of Linear Equations

You need to be able to

identify points that lie on the graph of a line.

What is the *x*-intercept of the graph of $5x - 4y + 10 = 0$?

To find the *x*-intercept, substitute 0 for *y* and solve for *x*.

Substitute 0 for *y*. $5x - 4(0) + 10 = 0$

Subtract 10 from both sides. $5x + 10 - 10 = 0 -$ ________

Divide both sides by 5. $5x \div 5 = -10 \div 5$

$x =$ ________

Which point on the graph of $x = 2y - 14$ has an *x*-coordinate of -8?

Substitute -8 for *x*. $-8 = 2y - 14$

Add 14 to both sides. $-8 + 14 = 2y - 14 + 14$

Divide both sides by 2. $6 \div 2 = 2y \div$ ________

$y =$ ________

Write the coordinates of the point. $(-8, 3)$

TEST PRACTICE 1

Which equation has a graph that contains the point $(-2, 3)$?

A $3x - 2y = 12$ **C** $2x - 3y = 12$

B $3x - 2y = -12$ **D** $2x - 3y = -12$

Hint: Which answer choice is true when $x = -2$ and $y = 3$?

CAHSEE Standards Mini-Review 9

Standards 1A6.0–1A9.0

You need to be able to

use the slope-intercept form to graph an equation.

An equation in the form $y = mx + b$ is in slope-intercept form. The slope is m and the y-intercept is b.

Change $2x + y = 1$ to slope-intercept form. Graph the equation.

Subtract $2x$ from both sides. $2x + y - 2x = 1 - 2x$

Write the slope-intercept form. $y = -2x + 1$

Identify the slope and y-intercept. slope $= -2$, y-intercept $=$ _____

Mark the y-intercept.

The slope is -2. Go over 1 unit and down 2 units to mark a second point.

Connect the points with a straight line.

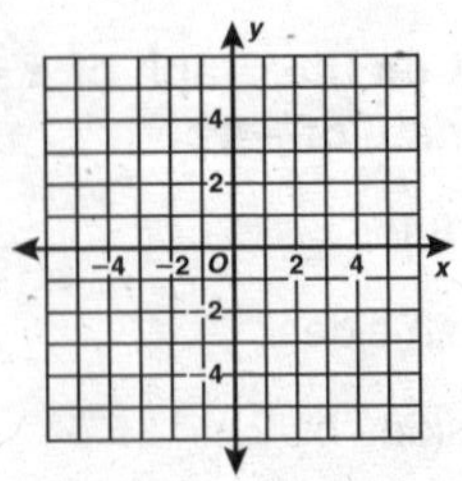

TEST PRACTICE 2

Graph this equation on the grid at the right. Then choose the y-intercept.

$2x - 3y = 3$

A -1 **C** -3

B -2 **D** 3

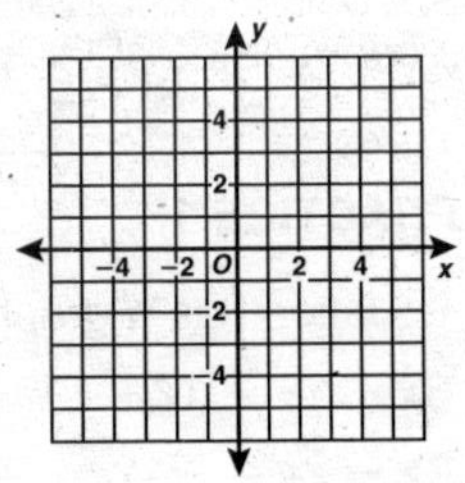

Hint: Any equation can be graphed from a table of ordered pairs.

CAHSEE Standards Mini-Review 9

Graphing and Systems of Linear Equations

You need to be able to

use substitution to solve a system of equations.

Use substitution to solve this system of equations. $\begin{cases} y = 7 - 2x \\ y = -5x + 1 \end{cases}$

Make sure both equations are solved for y. Start with either equation. Substitute the expression for y from the other equation, then solve for x.

Substitute $-5x + 1$ for y. $-5x + 1 = 7 - 2x$

Add $5x$ to both sides. $-5x + 1 + 5x = 7 - 2x +$ _____

Subtract 7 from both sides. $1 - 7 = 3x + 7 - 7$

Divide both sides by 3. $-6 \div 3 = 3x \div$ _____

$x =$ _____

Solve for the other variable using either original equation.

Substitute -2 for x in the first equation. $y = 7 - 2(-2)$

$y =$ _____

Write the solution as an ordered pair. $(-2, 11)$

TEST PRACTICE 7

What is the solution to this system of equations? $\begin{cases} y = 5x + 8 \\ 4x - y + 7 = 0 \end{cases}$

A $(-1, 3)$ **B** $(1, -3)$ **C** $(1, 13)$ **D** $(-9, -37)$

Hint: Solve the second equation for y. Then use substitution.

CAHSEE Standards Mini-Review 9

Standards 1A6.0–1A9.0

You need to be able to

use graphing to solve a system of equations.

Solve this system of equations by graphing. $\begin{cases} x + y + 2 = 0 \\ 2x - y + 1 = 0 \end{cases}$

Graph $x + y + 2 = 0$. First write it in slope-intercept form.

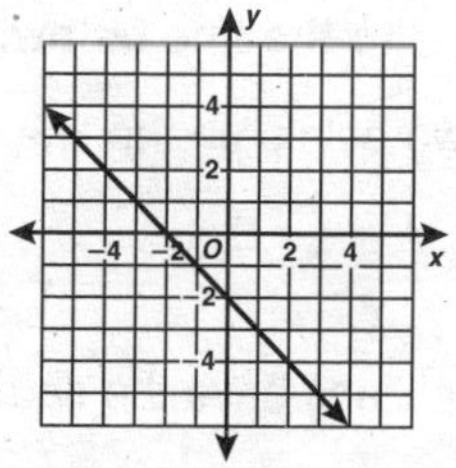

Graph $y =$ ____________

Graph $2x - y + 1 = 0$. First write it in slope-intercept form.

Graph $y =$ ____________

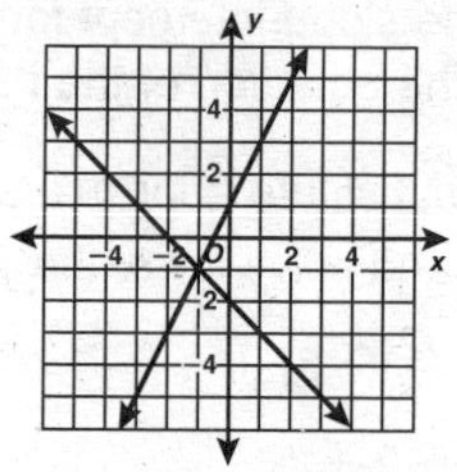

Write the coordinates of the point where the two lines intersect.

____

TEST PRACTICE 6

Graph this system on the grid at the right. Then choose the solution of the system.

$\begin{cases} y = 2 - x \\ x - 2y + 1 = 0 \end{cases}$

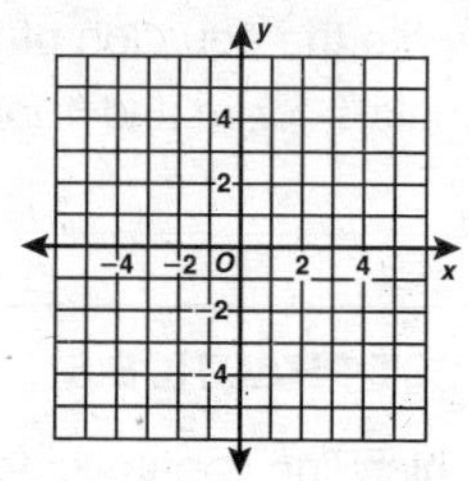

A $(0, \frac{1}{2})$ **C** $(1, 1)$

B $(\frac{1}{2}, 0)$ **D** $(0, 2)$

Hint: The slope-intercept form of the second equation is $y = \frac{1}{2}x + \frac{1}{2}$.

CAHSEE Standards Mini-Review 9

Graphing and Systems of Linear Equations

You need to be able to

graph the solution set of an inequality.

Graph the solution set of $x < 1 + y$.

Use a dotted line to graph the equation $x = 1 + y$. First write it in slope-intercept form.

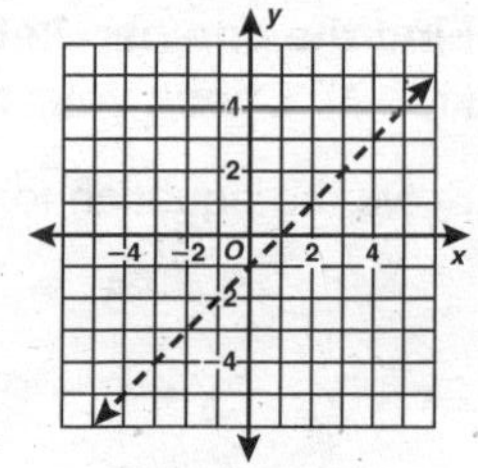

Subtract 1 from both sides.

$$x = 1 + y$$

$$x - 1 = 1 + y - 1$$

Graph $y =$ ____________

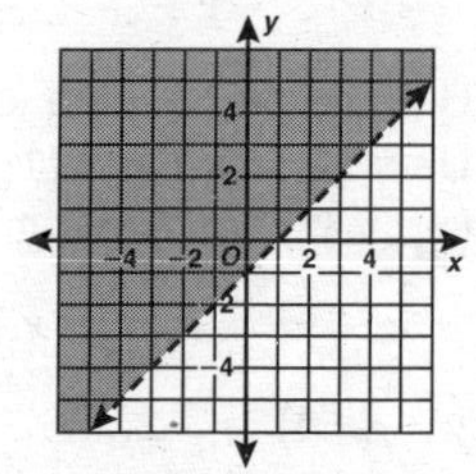

Test the point (0, 0) to find out if it is a solution of $x < 1 + y$.

$0 < 1 + 0$ is true. So, shade the region that contains (0, 0).

TEST PRACTICE 3

Which inequality is shown on the graph at the right?

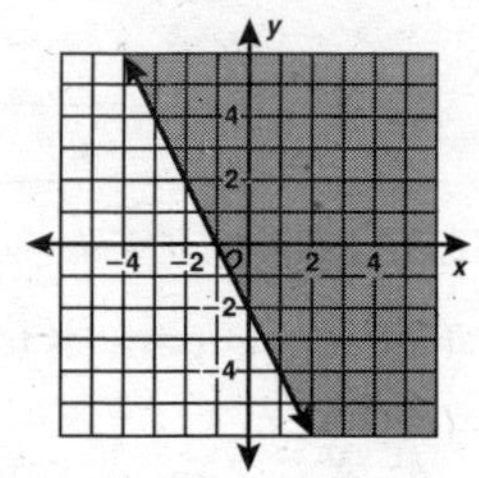

A $2x + y > 2$ **C** $2x + y \leq -2$

B $2x + y > -2$ **D** $2x + y \geq -2$

Hint: Use the symbol $\geq$ or $\leq$ when the graph shows a solid line.

CAHSEE Standards Mini-Review 9

Standards 1A6.0–1A9.0

You need to be able to

find an equation of a line parallel to a given line that passes through a given point.

Find the line parallel to $2x + y = 3$ that passes through the point (1, −4). Check your answer by graphing both lines.

Solve the equation for y to find the slope. Subtract $2x$ from both sides.

$$2x + y = 3$$

$$2x + y - 2x = 3 - 2x$$

$$y = \_\_\_\_\_\_\_\_$$

Use the point-slope formula.
Use −2 for m and (1, −4) for (x_1, y_1). Then solve for y.

$$y - y_1 = m(x - x_1)$$

$$y - (-4) = -2(x - 1)$$

$$y + 4 = -2x + 2$$

$$y = \_\_\_\_\_\_\_\_$$

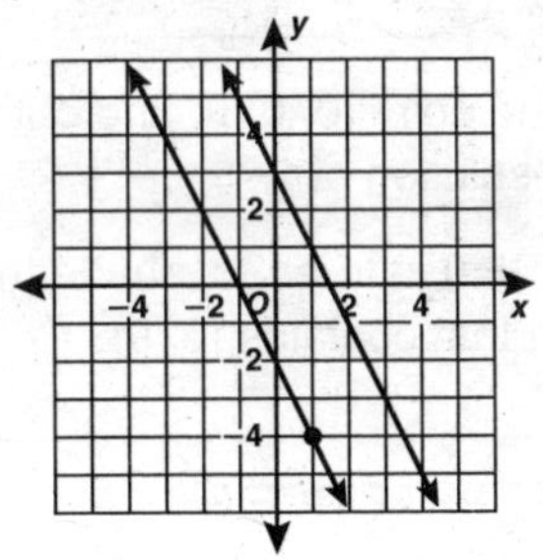

TEST PRACTICE 4

Which line is parallel to $y = \frac{1}{3}x - 2$ and contains the point (3, 2)?

A $y = 3x + 1$ **C** $y = \frac{1}{3}x + 1$

B $y = 3x + 2$ **D** $y = \frac{1}{3}x + 2$

Hint: Two lines are parallel if their slopes are equal but their y-intercepts are different.

Page 4

CAHSEE Standards Mini-Review 9

Graphing and Systems of Linear Equations

You need to be able to

find an equation of a line parallel to a given line that passes through a given point.

Find the equation of the line that contains the point (−3, −1) and is parallel to the line below.

Use two points on the line shown to calculate the slope of the line shown.

$$\frac{\text{rise}}{\text{run}} = \frac{1 - (-1)}{5 - 2} = \_\_\_\_\_\_\_\_$$

Parallel lines have the same slope.

Use the slope-intercept formula to write the equation of the parallel line.

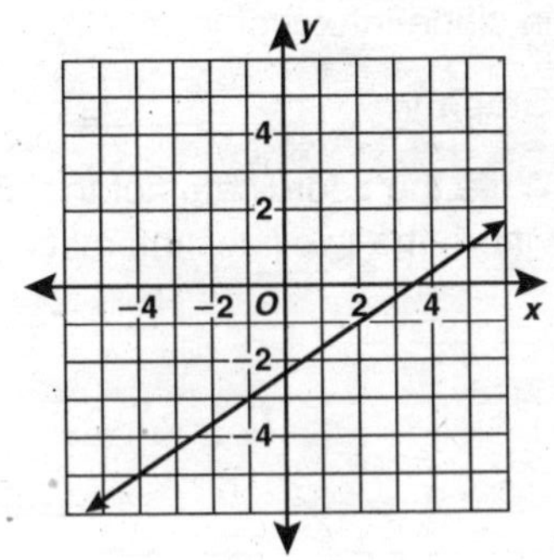

First, substitute $\frac{2}{3}$ for m and (−3, −1) for (x, y) and solve for b, the y-intercept.

$$-1 = (\tfrac{2}{3})(-3) + b$$

$$-1 = -2 + b$$

$$\_\_\_ = b$$

To write the equation of the new line, substitute into the slope-intercept formula $\frac{2}{3}$ for m and 1 for b.

$$y = \_\_\_\_\_\_\_\_$$

TEST PRACTICE 5

Which line is parallel to $y = \frac{x}{4} + 2$?

A $y = \frac{x}{4} - 1$ **B** $y = -\frac{x}{4} - 1$ **C** $y = 4x - 1$ **D** $y = -4x - 1$

Hint: Parallel lines have the same slope but different y-intercepts.

Page 5

CAHSEE Standards Mini-Review 10

Standards 1A10.0–1A11.0, 1A14.0, 1A19.0–1A23.0

You need to be able to

graph a quadratic function and identify any x-intercepts.

Find the x-intercepts of the graph of $y = x^2 - 2x + 1$.

Set y equal to 0. $0 = x^2 - 2x + 1$

Factor the right side. $0 = (x - 1)(x - 1)$

Set each factor equal to 0. $x - 1 = 0$ $x - 1 =$ ______

Solve each equation. $x = 1$ $x =$ ______

Both roots of the equation equal 1. There is just one x-intercept at (1, 0).

Graph $y = x^2 - 2x + 1$.

Make a table of ordered pairs. Plot the points and connect them to make a bowl-shaped curve called a *parabola*.

Complete the table and the graph.

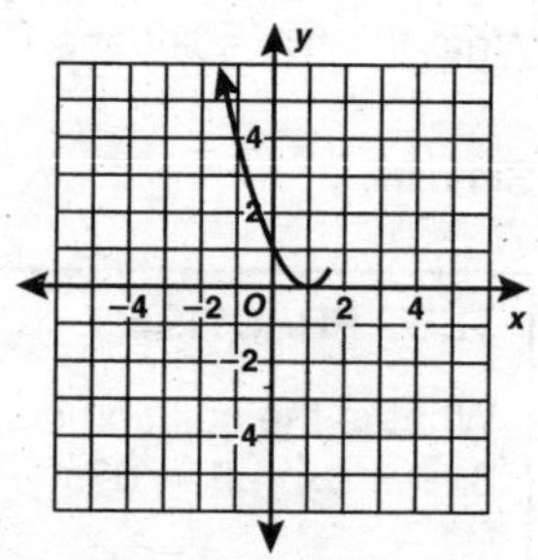

x	−1	0	1	2	3
y	4	1	0		

TEST PRACTICE 8

What are the x-intercepts of $y = x^2 + 5x + 4$?

A −1, −4 **B** −1, 4 **C** 1, −4 **D** 1, 4

Hint: If the quadratic equation has no real roots, the graph of the quadratic function will have no x-intercepts. For example, try graphing $y = x^2 - 6x + 11$.

CAHSEE Standards Mini-Review 10

Quadratics and Polynomials

You need to be able to

add and subtract polynomials.

$(2m^2 - 4) + (5m^2 - 8m) + (1 - m) =$

Add the m^2 terms. $2m^2 + 5m^2 =$ ______

Add the m terms. $-8m + (-m) = -9m$

Add the constant terms. $-4 + 1 =$ ______

Write the terms in the sum. $7m^2 - 9m - 3$

$(3a^2 - 7) - (8a^2 + a - 11) =$

To subtract, change each term to its opposite then add coefficients.

Write the second polynomial under the first. Line up like terms, those with variables of the same degree.

$$\begin{array}{r} 3a^2 \qquad - 7 \\ \underline{8a^2 + a - 11} \end{array}$$

Write the opposite of each term being subtracted.

$$\begin{array}{r} 3a^2 \qquad - 7 \\ \underline{-8a^2 - a + 11} \end{array}$$

Add the terms. $-5a^2 - a +$ ______

TEST PRACTICE 1

$(3x + 2xy - y) - (4y - 11xy) =$

A $3x + 13xy - 5y$ **C** $3x + 13xy - 3y$

B $3x - 9xy - 5y$ **D** $3x - 10xy - 3y$

Hint: What is the opposite of $(4y - 11xy)$?

CAHSEE Standards Mini-Review 10

Standards 1A10.0–1A11.0, 1A14.0, 1A19.0–1A23.0

You need to be able to

multiply polynomials.

$(4a^2b) \cdot (-10a^3b^2) =$

Multiply the coefficients. $4 \cdot -10 =$ ______

Multiply the *a* terms.
Add the exponents 2 and 3. $a^2 \cdot a^3 = a^5$

Multiply the *b* terms.
Add the exponents 1 and 2. $b \cdot b^2 =$ ______

Write the product. $-40a^5b^3$

$-5y\,(2y^2 - 7) =$

Use the Distributive Property to multiply a monomial times a polynomial.

Multiply $-5y$ times $2y^2$. $-5y \cdot 2y^2 = -10$ ______

Multiply $-5y$ times -7. $-5y \cdot -7 = 35y$

Write the product. $-10y^3 +$ ______

TEST PRACTICE 2

$-2c^2d\,(3c - 5d^2)$

A $-6c^2d^3 + 10c^3d$ **C** $4c^6d^3$

B $-6c^3d + 10c^2d^3$ **D** $4c^5d^4$

Hint: Remember, to multiply two exponential expressions with the same base, add the exponents.

CAHSEE Standards Mini-Review 10

Quadratics and Polynomials

You need to be able to

use the quadratic formula to solve a quadratic equation.

Solve $25x^2 - 15x - 28 = 0$ using the quadratic formula.

Write the quadratic formula. $x = \dfrac{-b + \sqrt{b^2 - 4ac}}{2a}$

Identify the values for *a, b, c.* $a =$ ______, $b = -15$, $c = -28$

Substitute the values. $x = \dfrac{-(-15) \pm \sqrt{(-15)^2 - 4(25)(-28)}}{2(25)}$

Simplify. $x = \dfrac{15 \pm \sqrt{225 + 2800}}{50} = \dfrac{15 \pm \sqrt{3025}}{50} = \dfrac{15 \pm 55}{50}$

Write the two solutions. $x = \dfrac{15 + 55}{50}$ $x = \dfrac{15 - 55}{50}$

Simplify. $x = \dfrac{7}{5}$ $x =$ ______

TEST PRACTICE 7

What is the solution set of the quadratic equation $8x^2 + 6x - 5 = 0$?

A $\{-8, 5\}$ **C** $\left\{-\frac{5}{4}, \frac{1}{2}\right\}$

B $\{8, -5\}$ **D** $\left\{\frac{5}{4}, -\frac{1}{2}\right\}$

Hint: When the coefficient of the x^2 term is not 1, the quadratic formula is usually easier than factoring.

CAHSEE Standards Mini-Review 10

Standards 1A10.0–1A11.0, 1A14.0, 1A19.0–1A23.0

You need to be able to

use factoring to solve a quadratic equation.

Solve $n^2 - 3n = 10$ by factoring.

Subtract 10 from both sides.	$n^2 - 3n - 10 = 10 - 10$	
	$n^2 - 3n - 10 =$ ______	
Factor the left side of the equation.	$(n + 2)(n - 5) = 0$	
Set each factor equal to 0.	$n + 2 = 0$	$n - 5 = 0$
Solve.	$n = -2$	$n =$ ______

Use set brackets { } to write the two solutions. ________

Solve $9q^2 - 25 = 0$ by factoring.

The left side is the difference of two squares.

Factor the left side of the equation.	$(3q + 5)(3q - 5) = 0$	
Set each factor equal to 0.	$3q + 5 = 0$	$3q - 5 = 0$
Solve.	$q = -\frac{5}{3}$	$q =$ ______

TEST PRACTICE 6

What is the solution set of the quadratic equation $2y^2 + 11y - 6 = 0$?

A $\left\{\frac{1}{2}, 6\right\}$ **B** $\left\{-\frac{1}{2}, -6\right\}$ **C** $\left\{-\frac{1}{2}, 6\right\}$ **D** $\left\{\frac{1}{2}, -6\right\}$

Hint: This strategy works because of the Zero-Product Property: For all real numbers *a* and *b*, if the product *ab* equals 0, then either $a = 0$ or $b = 0$.

CAHSEE Standards Mini-Review 10

Quadratics and Polynomials

You need to be able to

use the greatest common factor to factor a polynomial.

What is the greatest common factor of the terms in this polynomial?

$24n^3 - 18n^2 + 12n$

Find the greatest common factor of the coefficients 24, 18, and 12.	GCF(24, 18, 12) = 6
Find the greatest common factor of the variable terms n^3, n^2, and *n*.	GCF(n^3, n^2, n) = n
Multiply.	GCF = $6 \cdot n$ = ______

Factor $20x^2 - 12x^3$.

Find the GCF of the terms.	GCF($20x^2$, $12x^3$) = $4x^2$
Divide each term by the GCF.	$20x^2 \div 4x^2 = 5$
	$12x^3 \div 4x^2 =$ ______
Write the product.	$4x^2$ (______ $- 3x$)

TEST PRACTICE 3

Factor $24k + 4k^3$ completely.

A $6(4k + k^2)$ **C** $4k(6 + k^2)$

B $2k(12 + 2k^2)$ **D** $4(6k + k^3)$

Hint: Check that the factors in your answer have no more common factors. Choice B is not correct because 12 and $2k^2$ have the common factor 2.

CAHSEE Standards Mini-Review 10

Standards 1A10.0–1A11.0, 1A14.0, 1A19.0–1A23.0

You need to be able to

recognize the difference of two squares or the square of a binomial.

Watch for these factoring patterns.

Difference of Two Squares	Square of Binomial
$x^2 - 1 = (x + 1)(x - 1)$	$x^2 + 2Ax + A^2 = (x + A)(x + A)$
$x^2 - y^2 = (x + y)(x - y)$	$x^2 - 2Ax + A^2 = (x - A)(x - A)$

Factor $25w^2 - \frac{1}{4}$.

Check that each term is the square of a monomial. $25w^2 = (5w)^2$ $\quad \frac{1}{4} = \left(\frac{1}{2}\right)^2$

Write the product. $\left(5w + \frac{1}{2}\right)(5w - \_\_\_\_)$

Factor $4a^2 - 12ab + 9b^2$.

Check that the first and last terms of the trinomial are squares. If the middle term is negative, use the opposite of the second monomial.

Write the terms as squares. $4a^2 = (2a)^2$ $\quad 9b^2 = (-3b)^2$

Check that twice the product of the monomials equals the middle term. $2 \cdot 2a \cdot -3b = \_\_\_\_$

Write the square. $(2a - 3b)^2$

TEST PRACTICE 4

What is the factored form of $4n^2 + 4n + 1$?

A $(2n - 1)^2$ **B** $(2n + 1)^2$ **C** $(4n - 1)^2$ **D** $(4n + 1)^2$

Hint: Check your factors by multiplying.

CAHSEE Standards Mini-Review 10

Quadratics and Polynomials

You need to be able to

factor trinomials.

Factor $x^2 - 10 - 3x$.

Transform the polynomial to the form $x^2 + bx + c$. $\quad x^2 - 3x - 10$

List pairs of factors of the third term. Find the sum of each pair of factors.

Factors	Sum
1, −10	−9
−1, 10	9
−2, 5	3
2, −5	−3

Choose the factors with the sum *b*. In this example, the sum must be −3. The two factors with that sum are 2 and ____.

Use the two factors to write two binomials. $(x + 2)(x - \_\_\_\_)$

Factor $3c^3 - 6c^2 - 24c$.

The greatest common factor is $3c$. Divide each term by $3c$. $\quad 3c(c^2 - 2c - 8)$

Use the method above to factor the trinomial $c^2 - 2c - 8$. $\quad 3c(c - 4)(\_\_\_\_)$

TEST PRACTICE 5

Which is a factor of $h^2 - 8h - 20$?

A $h + 2$ **B** $h - 2$ **C** $h + 10$ **D** $h - 20$

Hint: Which pair of factors of −20 has a sum of −8?

CAHSEE Standards Mini-Review 11

Standards 1A12.0–1A13.0, 1A15.0–1A18.0

You need to be able to

find the domain and range of a relation or function.

The range of a relation can be determined by the domain and a description or rule. The rule is often given by an equation.

What is the range of this function?

Domain = {positive integers} $y = 2x$

List the first 5 members of the domain and range.

$D = \{1, 2, 3, 4, 5, \ldots\}$

$R = \{2, 4, 6, 8, 10, \ldots\}$

The range is the set of positive even integers.

What is the domain of the relation shown on the graph at the right?

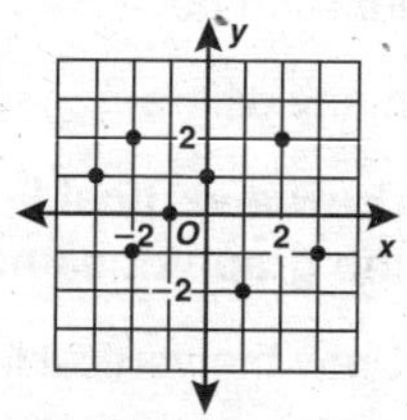

List the x-coordinates of the points.

$D = \{-3, -2, -1, 0, 1,$ ________ $\}$

TEST PRACTICE 8

What is the range of this function?

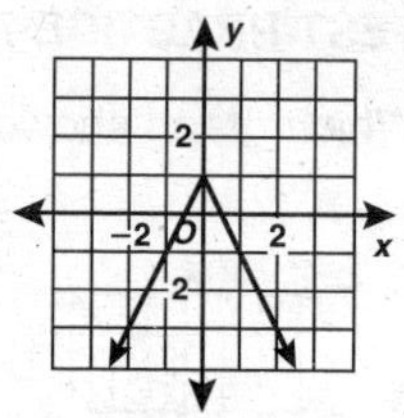

A all integers

B integers ≤ 1

C real numbers ≤ 1

D all real numbers

Hint: If the domain isn't specified, assume it is all real numbers for which the function rule is meaningful.

CAHSEE Standards Mini-Review 11

Functions and Rational Expressions

You need to be able to

simplify fractions with polynomials in the numerator and denominator.

To simplify a rational expression, divide numerator and denominator by the greatest common factor.

What is $\dfrac{2ab^2 + 10b^2}{a^2b + 2ab - 15b}$ reduced to lowest terms?

Factor the numerator. $\dfrac{2b^2(a + 5)}{a^2b + 2ab - 15b}$

Factor the denominator. $\dfrac{2b^2(a + 5)}{b(a - 3)(a + 5)}$

Find the greatest common factor. GCF $= b(a + 5)$

Divide both terms by $(a + 5)$. $\dfrac{2b^2\cancel{(a + 5)}}{b(a - 3)\cancel{(a + 5)}} =$ ________

Simplify $\dfrac{b^2}{b}$. $\dfrac{2b^2}{b(a - 3)} = \dfrac{\quad}{a - 3}$

TEST PRACTICE 1

Simplify $\dfrac{w^2 - 9}{2w^2 - 9w + 9}$.

A $\dfrac{w - 9}{2w}$

B $\dfrac{-8}{11 - 9w}$

C $\dfrac{w + 3}{2w - 3}$

D $\dfrac{w + 3}{2(w - 3)}$

Hint: One factor of the denominator is $2w - 3$. What is the other factor?

CAHSEE Standards Mini-Review 11

Standards 1A12.0–1A13.0, 1A15.0–1A18.0

You need to be able to

multiply and divide rational expressions.

To divide rational expressions, multiply by the reciprocal of the divisor. Create the reciprocal by interchanging numerator and denominator.

Find the quotient. $\frac{d}{d+2} \div \frac{30}{10+5d}$

Write the reciprocal of the divisor. $\frac{d}{d+2} \cdot \frac{10+5d}{30}$

Factor numerators and denominators. $\frac{d}{d+2} \cdot$ ________

Multiply. Leave terms in factored form. $\frac{5d(2+d)}{30(d+2)}$

Divide by the common factor $(d + 2)$. $\frac{5d\cancel{(2+d)}}{30\cancel{(d+2)}} =$ ________

Divide by the common factor 5. $\frac{5d}{30} =$ ________

TEST PRACTICE 2

$\frac{n-3}{8} \cdot \frac{4n-10}{n^2+n-12} =$

A $\frac{n-5}{2(n+4)}$

B $\frac{2n-5}{4(n+4)}$

C $\frac{3n-8}{(n-3)(n+4)}$

D $\frac{5n-13}{8(n^2+n-12)}$

Hint: Start by factoring the terms of the second factor.

CAHSEE Standards Mini-Review 11

Functions and Rational Expressions

You need to be able to

identify a function from a graph.

A relation can be shown on a graph. If a vertical line intersects the graph in more than one point, the relation is not a function.

Does each graph show a function? Explain.

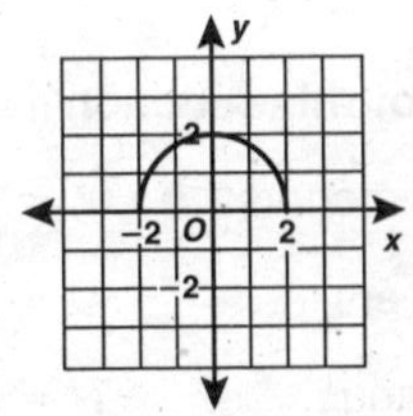

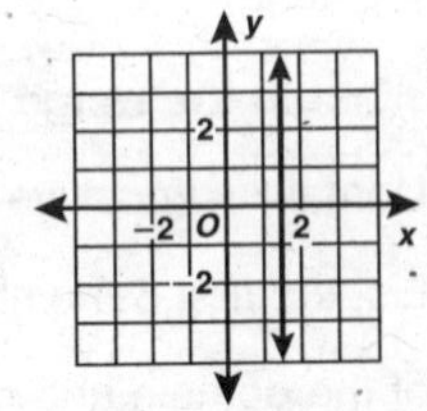

The left graph shows a function because each *x*-value is paired with a different *y*-value.

The right graph does not show a function because *x*-value 1.5 is paired with many different ________.

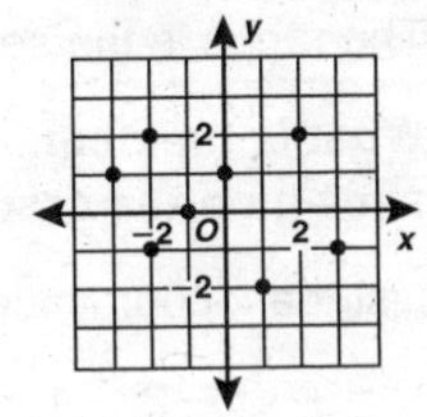

Draw a vertical line to show that the graph does not show a function.

Draw the vertical line through −2.

TEST PRACTICE 7

Which graph shows a function?

A

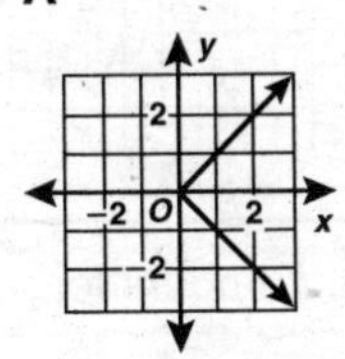

B

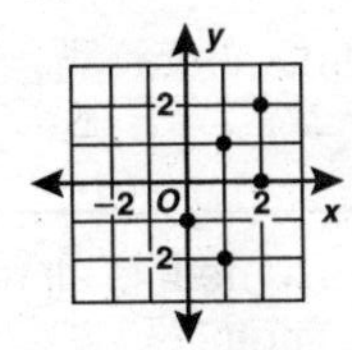

C

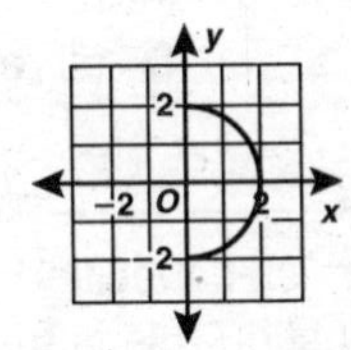

D

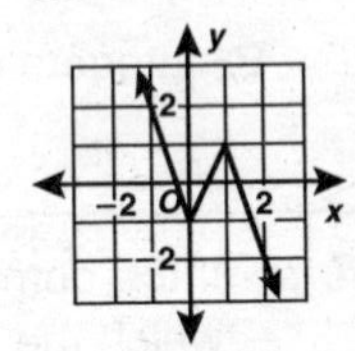

Hint: Three of the graphs show one *x*-value paired with two different *y*-values.

CAHSEE Standards Mini-Review 11

Standards 1A12.0–1A13.0, 1A15.0–1A18.0

You need to be able to

identify a function from a set of ordered pairs.

A relation is any set of ordered pairs. The set of first coordinates in the ordered pairs is called the domain of the relation. The set of second coordinates is the range.

A function is a relation that pairs each element of the domain with exactly one element of the range.

Describe the relation shown in this table of ordered pairs. Is the relation a function? Give a reason for your answer.

Domain	1	1	4	4	9	9
Range	1	−1	2	−2	3	−3

The relation pairs each element in the domain with its positive and negative square root. Complete these ordered pairs to show other members of this function.

(16, 4) (16, ______) (25, −5)(25, ______)

The relation is not a function because, for example, 4 is paired with two different numbers.

TEST PRACTICE 6

Which relation is a function?

A {(1, 2), (1, 3), (1, 4), (1, 5)} **C** {(1, 2), (2, 3), (1, 4), (3, 5)}

B {(2, 1), (3, 1), (4, 1), (5, 1)} **D** {(8, 1), (2, 4), (4, 2), (8, 8)}

Hint: Choice D can be eliminated because 8 is paired with more than one number. What other choices can you eliminate?

CAHSEE Standards Mini-Review 11

Functions and Rational Expressions

You need to be able to

add and subtract rational expressions.

Find the difference. $\dfrac{3x+5}{x-2} - \dfrac{1+x}{3x-6}$

Factor numerators and denominators. $\dfrac{3x+5}{x-2} - \dfrac{1+x}{\_\_\_\_}$

Multiply the first fraction by $\frac{3}{3}$ to make the denominators the same. $\dfrac{3(3x+5)}{3(x-2)} - \dfrac{1+x}{3(x-2)}$

Expand the first numerator. $\dfrac{\_\_\_\_}{3(x-2)} - \dfrac{1+x}{3(x-2)}$

Find the opposite of the second numerator. Subtract like terms and simplify. Expand the denominator.

$$\frac{9x+15-(1+x)}{3(x-2)} = \frac{9x+15-1-x}{3(x-2)} = \frac{8x+14}{3(x-2)} = \frac{8x+14}{3x-6}$$

TEST PRACTICE 3

$\dfrac{10-5a}{a^2-4} + \dfrac{3}{a+2} =$

A $\dfrac{-2}{a+2}$ **C** $\dfrac{10-2a}{a^2-4}$

B $\dfrac{-2}{(a+2)^2}$ **D** $\dfrac{10-8a}{a^2-4}$

Hint: Simplify the first addend and you will have two expressions with the same denominator.

CAHSEE Standards Mini-Review 11

Standards 1A12.0–1A13.0, 1A15.0–1A18.0

You need to be able to

apply algebraic techniques to solve work problems.

A carpet layer can complete a standard size living room in 5 hours. It takes a less experienced worker 7 hours to do the same job. How long will it take for the job if both people work together?

Use h for the total number of hours. The first worker's rate of work is $\frac{1}{5}$ of the job per hour. In h hours, this worker can do $\frac{h}{5}$ of the job. The second worker can do only $\frac{h}{7}$ of the job. Write an equation.

first worker's part of job	+	second worker's part of job	=	whole job
$\frac{h}{5}$	+	$\frac{h}{7}$	=	1

Multiply both sides by the LCD. $35 \cdot \frac{h}{5} + 35 \cdot \frac{h}{7} = 1 \cdot$ ______

Simplify. $7h + 5h = 35$

Combine like terms. ______ $= 35$

Divide both sides by 12. $h = 2\frac{11}{12}$

$= 2$ h 55 min

TEST PRACTICE 4

Rick can rake leaves twice as fast as his little sister. They worked together and finished the job in 8 hours. How long would it take Rick to do the job by himself?

A 4 h **B** 6 h **C** 12 h **D** 16 h

Hint: Rick's sister's part of the job is $\frac{1}{2x} \cdot 8$. What is Rick's part?

CAHSEE Standards Mini-Review 11

Functions and Rational Expressions

You need to be able to

apply algebraic techniques to solve percent mixture problems.

Brand A of a trail mix is 65% nuts. Brand B is 40% nuts. Henry mixed some of each brand together and created a 90-lb mixture that was 55% nuts. How much of Brand A did Henry use?

To get the final mixture, Henry added some of Brand A to some of Brand B. Write an equation to show the pounds of nuts in each part.

lb nuts in Brand A part	+	lb nuts in Brand B part	=	lb nuts in final mixture
$65\% \cdot x$	+	$40\% \cdot (90 - x)$	=	$55\% \cdot 90$ lb

Convert the percents to decimals. $0.65x + 0.4(90 - x) = 0.55 \cdot 90$

Simplify both sides. $0.65x + 36 - 0.4x =$ ______

Combine like terms. $0.25x + 36 = 49.5$

Subtract 36 from both sides. $0.25x =$ ______

Divide both sides by 0.25. $x =$ ______

Henry used ______ pounds of Brand A to make his mixture.

TEST PRACTICE 5

A 10-gallon container of fruit punch is 20% water. How many gallons of a punch that is 8% water must be added to make a punch that is 12% water?

A 2 gal **B** 20 gal **C** 30 gal **D** 36 gal

Hint: Complete this equation, then solve for x.
$(20\% \cdot 10 \text{ gal}) + (8\% \cdot x \text{ gal}) = 12\%($______$)$ gal